Contents

Preface

I would like to start off with clichéd terms like heart is a wonderful piece of work, never stopping, pumping so much blood that puts the river Amazon to shame while being just about the size of one's pinky finger, beats a zillion times nonstop in your life, consumes the power of just about your night lamp, bla, bla.

However, if only it was true that the heart is such a marvelous and undefeatable organ this book would not exist. The purpose of this book is to present additional perspectives to professionals and students with the hope of extending the life of existing hearts and ultimately build hearts that are far superior to natural hearts.

The method of research for this book was mostly browsing through the Internet for related material to see if the ideas were even remotely feasible. I don't want to lie to you that I read all the material given in the references or that I have decades of academic or clinical experience. I also want to caution that this book is not a text book, not a research paper and not a clinical research report.

Let me not waste anymore words. Relish the ideas!

Dandelion
February 2016

Part 1 - Of the pump, by the pump and for the pump

Heart with its numerous valves, chambers, electrical circuits and vessels is a complicated organ and not easy to recreate with current technologies. This part explores alternative heart designs that may be easier to implement using biological or electromechanical building blocks. Let's deconstruct the features of the heart one by one. The word pump in this book refers to mechanical or biological and natural or artificial heart.

Left right pump separation

From the point of view of blood flow, the heart is actually two pumps joined into one organ. This conjoined architecture is difficult to replicate in artificial heart. It may be easier to create two artificial pumps instead of one

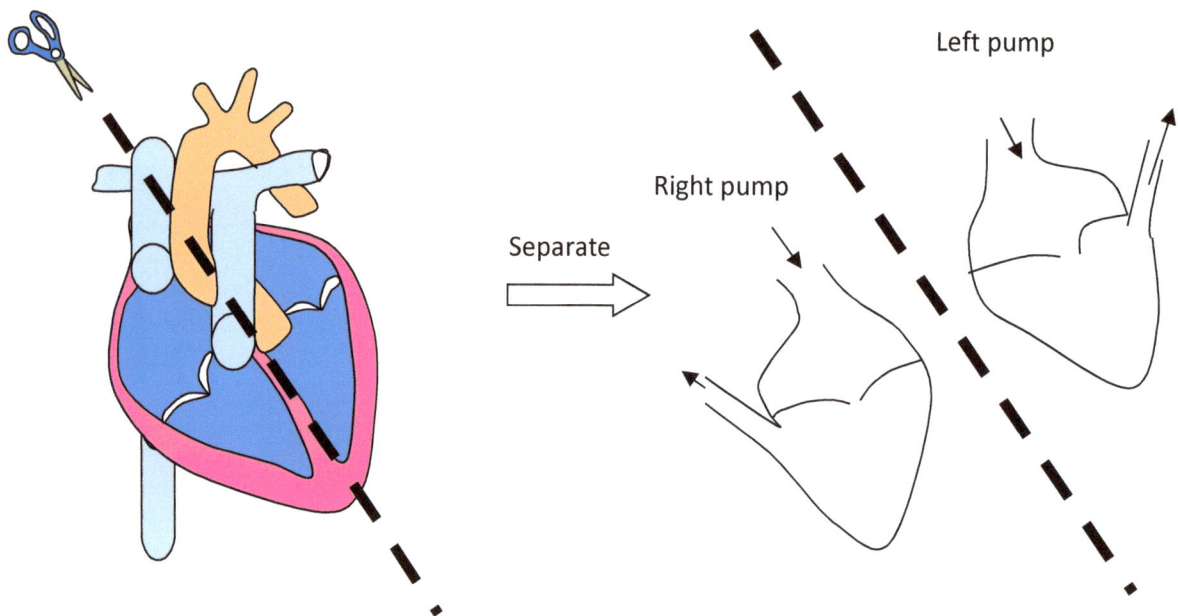

Atriumless pump

The atrium in human heart does indeed boost the output of the ventricle. But, it also adds a lot of complexity to the heart. To construct a simple heart, atrium may be unnecessary. There are many variations of the heart architecture across the animal kingdom. Fish for example has only one ventricle. The diagram below shows a single chamber pumping organ without atrium. Also, the outlet vessel is positioned in line with the inlet vessel to take advantage of the momentum of the incoming blood. In the natural heart, the incoming blood almost comes to a stop in the ventricle before being ejected. This probably wastes energy.

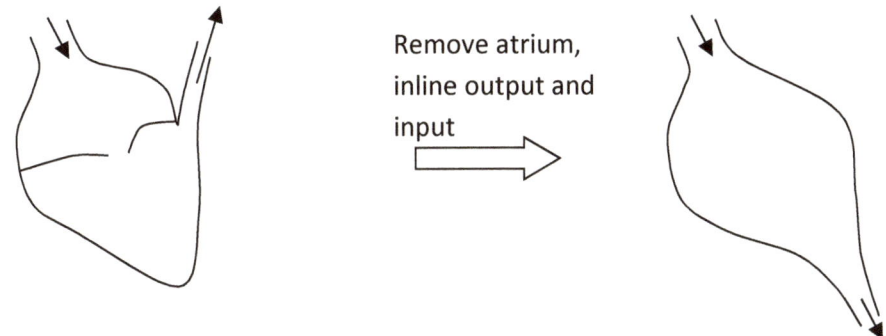

Remove atrium, inline output and input

Compound heart

The ventricle is a large chamber lined with muscle. Does a structure that needs to compress its contents need to be one large chamber? Actually not! Many small tube pumps to may get the same flow as one large pump while being suitable for manufacturing. Since the muscles of a single pump can be small, we can make it thin. A thin muscle is easily vascularized with blood vessels than a thicker muscle because vascularization of thick tissue is one of the main stumbling blocks to creating larger organs. The simplest structure we need for pumping is just a tube surrounded by muscle. With proper contraction rhythm the inlet and outlet valves are also not needed. Incidentally, this is how embryonic heart starts out. There are reports about artificial miniature heart that improves venous return.

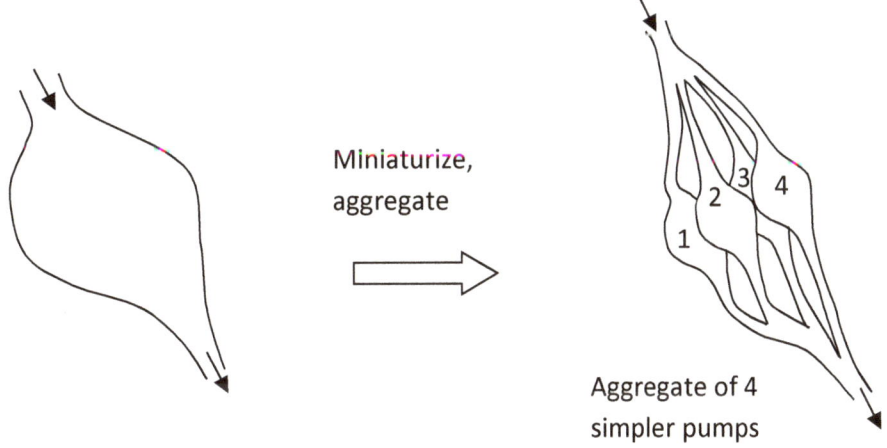

Miniaturize, aggregate

Aggregate of 4 simpler pumps

Coronary veins adulterating arteries

In the natural heart, the coronary circulation branch off very early in the arterial flow. The venous return is properly drained into the large veins going to the right heart. This gives good oxygen saturation in the arterial blood. But, this may be too complex to implement in a simplified artificial heart. The coronary venous return can be drained into the arterial output. This does mix oxygen poor blood with oxygen rich blood. The heart muscles take only about 5% of the blood circulation of the whole body. If we assume they take even up to 10% of the oxygen required by the whole body there is good 90% oxygen content remaining for the rest of the body. However, the problem that needs to be solved is that of vastly different arterial and venous pressure.

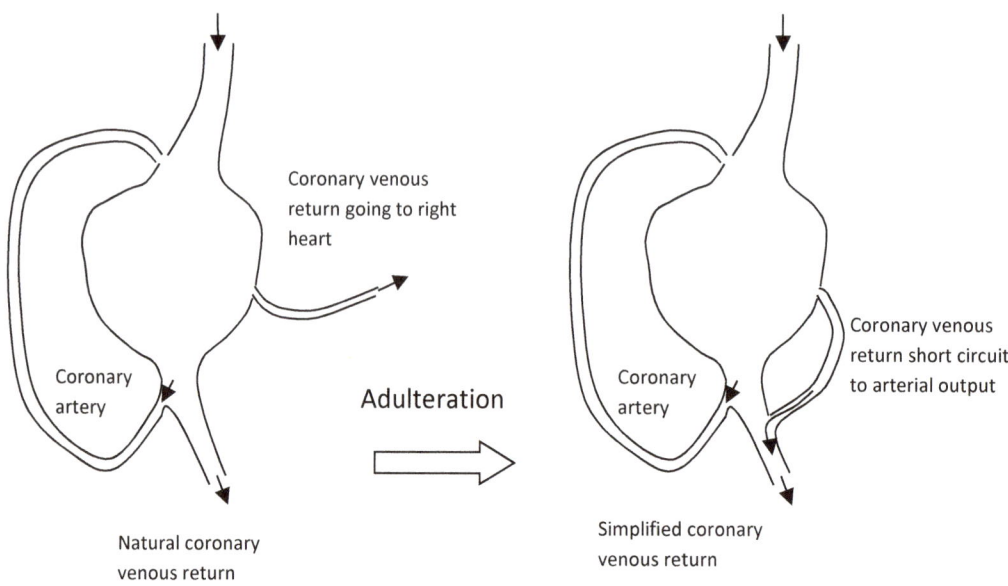

Coronary venous return going to right heart

Coronary artery

Adulteration

Natural coronary venous return

Coronary venous return short circuit to arterial output

Coronary artery

Simplified coronary venous return

Distributed Mini Hearts

The natural heart is lumped into one place. Alternative design can be made with many smaller pumps distributed throughout the arterial and venous network. The general advantage of a distributed system is that every unit in the system is simpler and the system is resilient to the loss of some units.

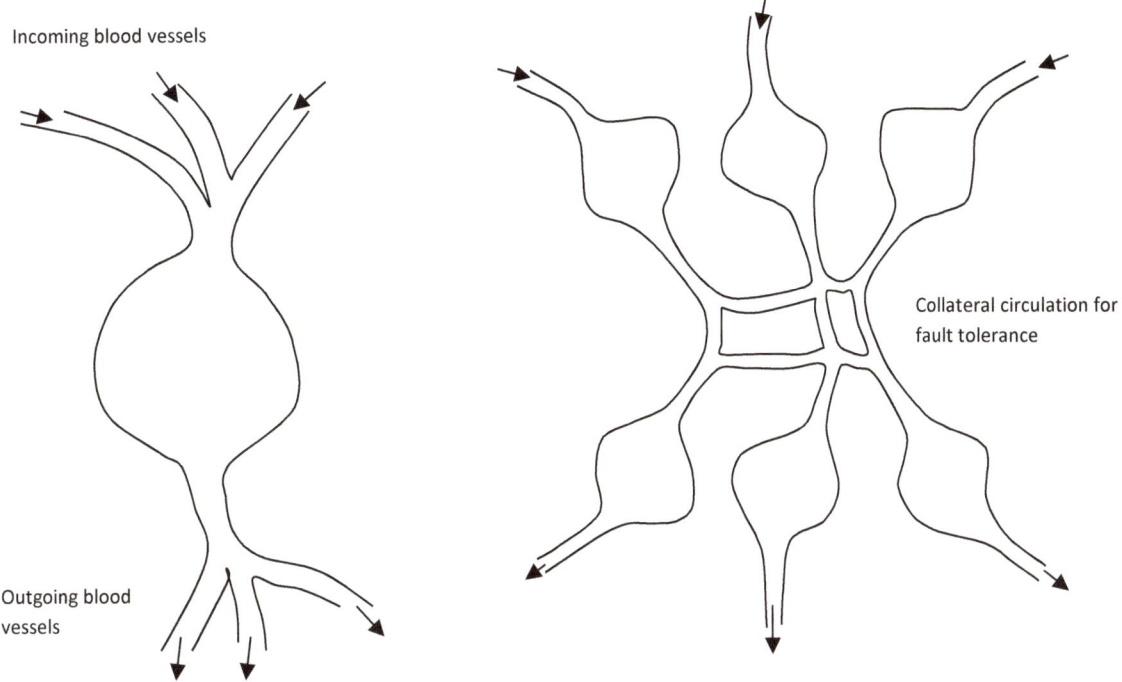

Incoming blood vessels

Collateral circulation for fault tolerance

Outgoing blood vessels

Simplified diagram of natural heart to show its lumped nature

Proposed distributed heart with many smaller pumps

Multiple Aorta

In the natural heart bulk of the outflow is through the aorta. In a synthetic heart we need not replicate the same design. We can have two outputs, one going up to the head and the other going down to rest of the body.

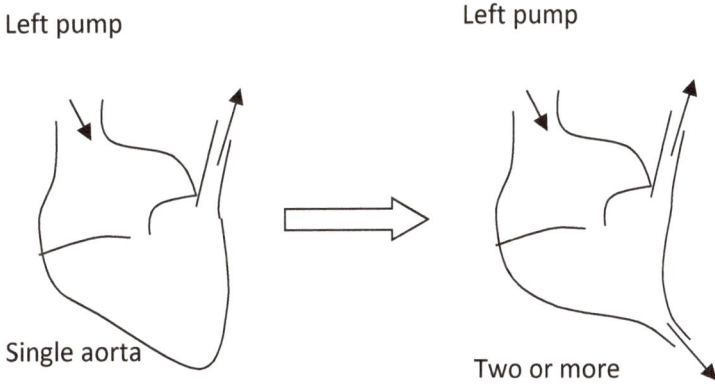

Left pump

Left pump

Single aorta

Two or more

Simple heart design for artificial synthesis

Combining some of the simplifications explained before, we can design a heart that has a good chance of working and has good ease of making. The heart will be formed of two pumps. The left pump will supply blood the body, the right pump with return the blood back to the lungs. A single chamber pump with shunt vessel avoids overpressure on the artery. The chamber fills with arterial blood as a regular chamber of the heart. Inlet and outlet valves prevent regurgitation. A pacing node sets the pump beating rhythm. A local coronary artery pumps the arterial blood back to the pump muscles. Native arterial flow and the pump flow are joined to produce higher cardiac output. For simplicity, venous drainage from pump muscles flow into outgoing artery. Heart rate can be increased to 300 beats/min to reduce chamber volume while maintaining good cardiac output. This simple design can be open loop, i.e., the heart output is not throttled by feedback signals. After the open loop design is achieved, a closed loop design can be attempted. It can be made to equalize the left and right heart output volumes and to maintain arterial-venous pressures and peripheral oxygen saturation within nominal range. The feedback can be done using pressure and oxygen sensors distributed throughout the conduction system. This simple heart may be realized with biological or electromechanical or hybrid approaches.

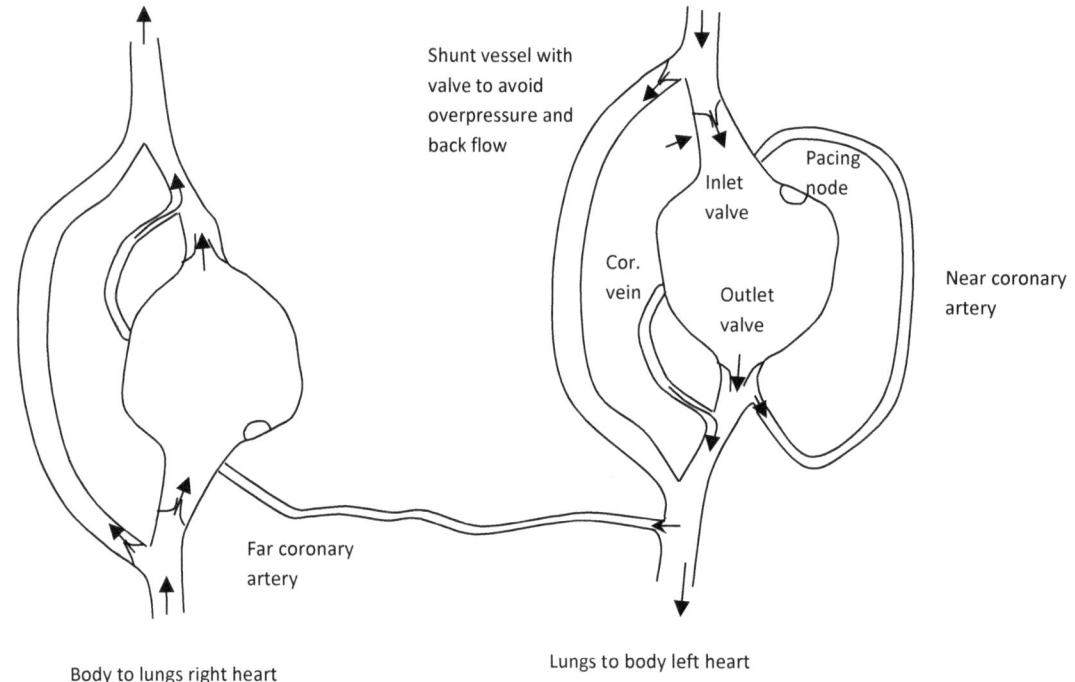

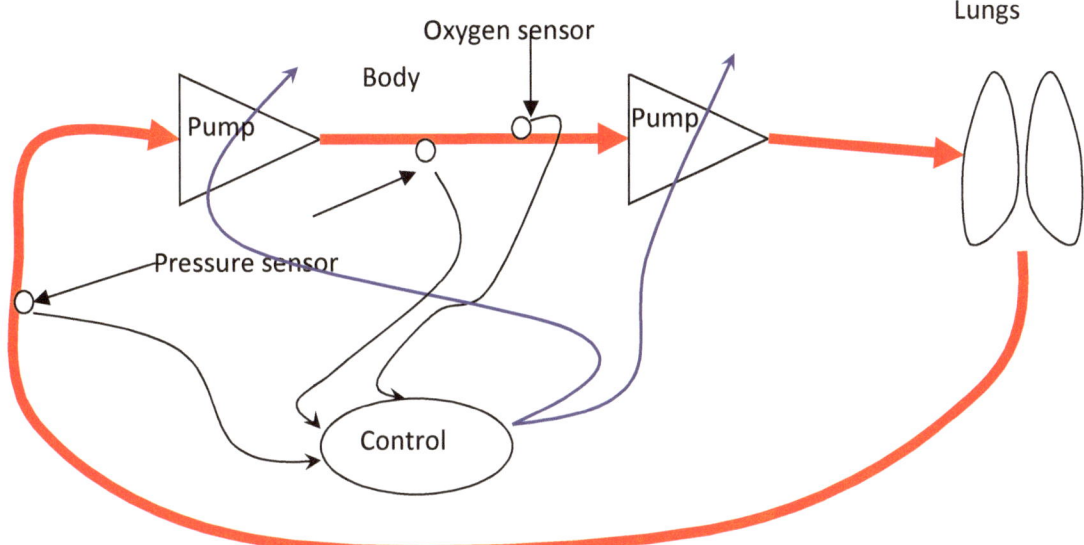

Whole body pump

During cardio pulmonary resuscitation (CPR), chest compression and relaxation provides the pumping power. When the chest relaxes after a compression no external power can be provided in this method. There are some techniques to improve pumping by taking advantage of the low pressure created in the thorax if inhalation is inhibited. Impedance threshold devices make use of this. Going one step further, we can attempt to use the peripheral blood to augment total circulation, for example, by adding peristaltic pump to the arms and legs. This whole body pump operation is mainly to compress the extremities when the thorax is relaxing, thus, augmenting the venous return. The compression can be synchronized to the chest compression using a wireless pressure sensor placed over the chest. The peripheral compression proceeds from the extremities towards the center resembling peristaltic pumping. The whole body pump can be automated by combining it with existing robotic CPR machines.

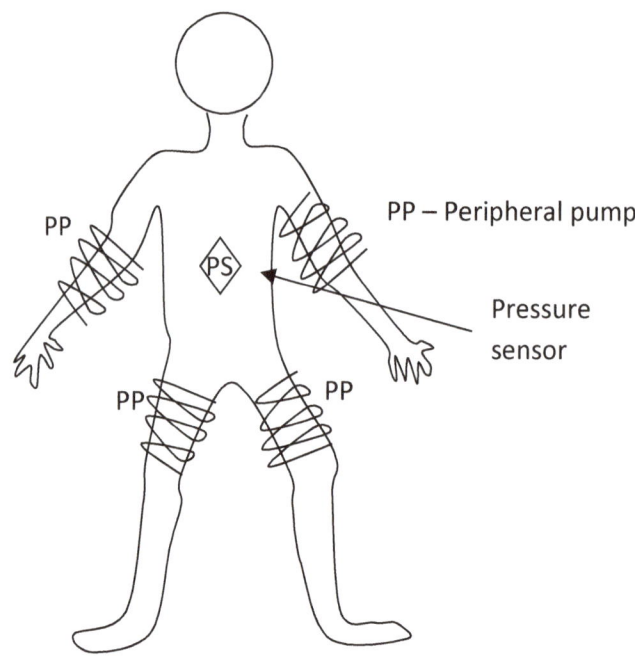

PP – Peripheral pump

Pressure sensor

Additional sensors for robotic CPR

Currently available robotic CPRs seem to use limited feedback from sensors. Some more information like oxygen saturation can be fed back.

Exotic heart designs

Of the different mechanical pumps, the heart seems to be similar to a diaphragm pump consisting of a compressible chamber and two valves. There are many more types of pumps that may be useful for an artificial heart. Some of these may offer different advantages in performance and ease of design.

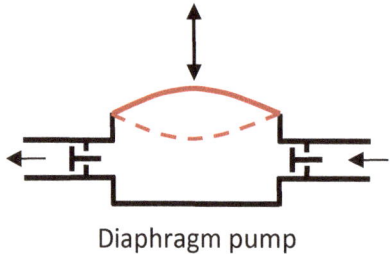

Diaphragm pump

Peristaltic pump

Peristaltic pump finds routine use in biology. The movement of food through the digestive system is by peristalsis. Coincidentally, heart-lung machines use peristaltic pumps too.

Peristaltic pump

Floating tissue biological pumps

Centrifugal pump is of particular interest as it has steady flow and possibly higher reliability. Centrifugal pump is also used in heart-lung machines. Centrifugal pumps are part of most of the modern Ventricular Assist Devices (VAD) and artificial mechanical hearts. Centrifugal pumps have one drawback that it needs full freedom rotating impeller. Macro biological systems almost never have a part that can rotate more than 360 degrees. The problem in a 360+ degree pump is that of supplying blood to the rotating part. How can a stationary blood vessel feed a continuously rotating part without getting spooled on it? To workaround this problem, we can consider a free floating impeller that is rotated by muscles attached to a chamber wall. Since the power for pumping is provided by muscles in the chamber wall and not the impeller itself, the impeller nutrient needs will be small to enable mere diffusion based circulation. Or the impeller circulation can be based on pressure gradients with valves to regulate the flow. This is akin to venous circulation that has no active pump but flows because of the pressure gradient and valves. The overall mechanism can be used to make not just centrifugal pump but axial pumps as well.

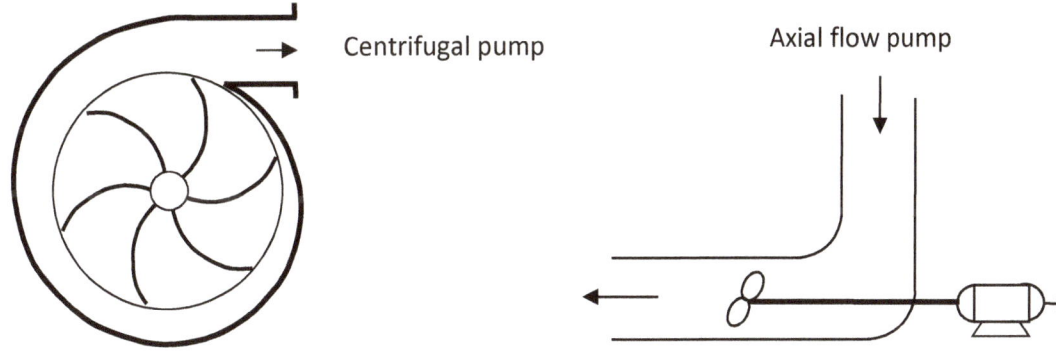

Centrifugal pump   Axial flow pump

Stroke 1: Push rotor fingers	Stroke 2: Tuck in stator fingers	Stroke 3: Push rotor fingers

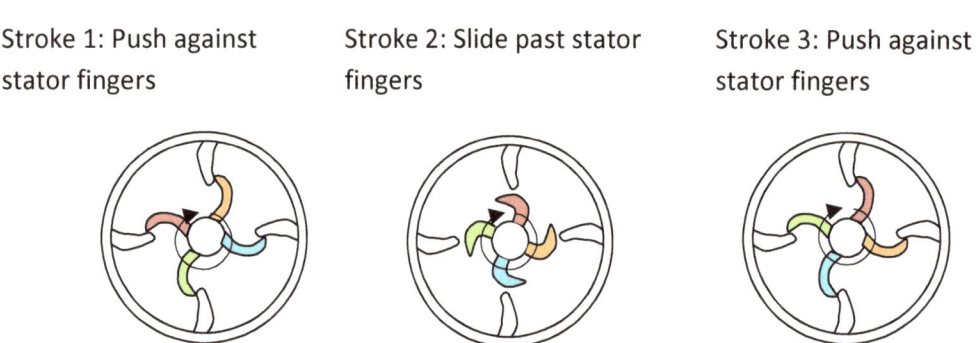

Rotation of floating impeller – Stator powered type

Stroke 1: Push against stator fingers	Stroke 2: Slide past stator fingers	Stroke 3: Push against stator fingers

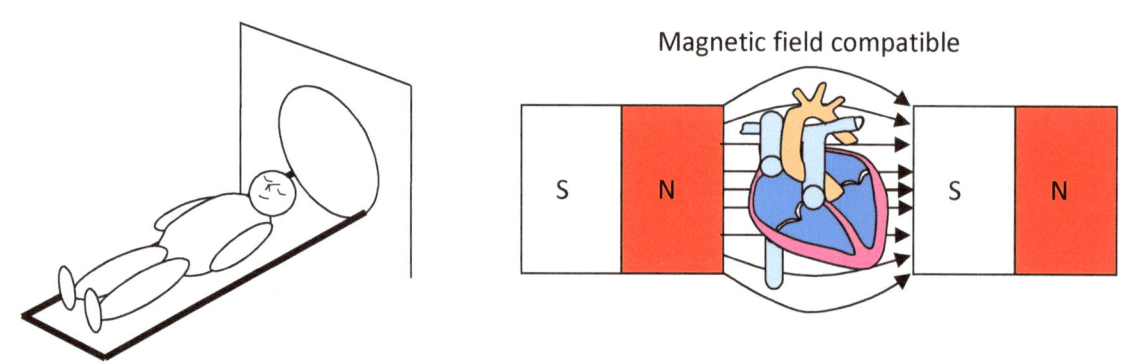

Rotation of floating impeller – Rotor powered type

MRI compatible artificial heart

Magnetic resonance imaging utilizes high magnetic fields. Most of the mechanical pumps also rely on creating magnetic fields and are incompatible with MRI. But, there are other pumping techniques that do not require magnetic fields or magnetic materials. These may be compatible with MRI machines. Piezo electric motor and electrostatic motor are in this category.

Magnetic field compatible

Blood acceleration tube

A hypothetical tube that accelerates blood as it passes through can be used to make artificial hearts. An electrostatic fluid accelerator can do this job but needs to ionize the fluid. Ionizing blood may have disastrous consequences. Another possibility could be to use hemoglobin's magnetic moment to accelerate inside the tube. The main advantage of the blood acceleration tube, if one can be made, will be its non contact nature reducing hemolysis and avoiding clotting.

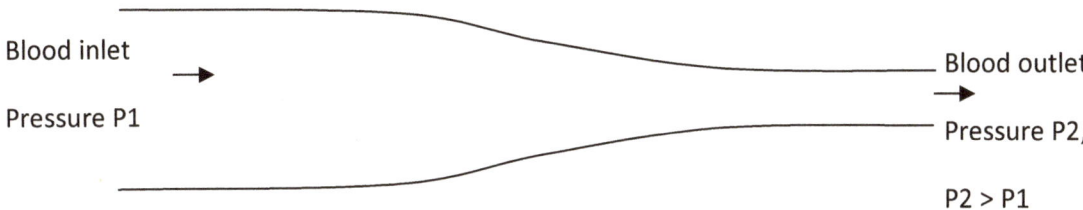

Blood inlet

Pressure P1

Blood outlet

Pressure P2,

P2 > P1

Part 2 - Stop conducting or I will burn you! – Arrhythmia and ablation

This part deals with heart conduction, abnormalities in conduction and some of the treatments used for conduction abnormalities. Just for recap, the heart muscle contractions are sequenced by pacing circuits in sino-atrial node and atrioventricular node. These pacing circuits supply electrical signals that travel through the heart muscle initiating contraction in specific sequence that gives required blood output. Arrhythmias occur when these circuits malfunction.

Permanent ion channel blocker

Ablation of abnormal conduction paths in the heart blocks the abnormal rhythms from happening. But this also destroys cardiomyocytes in these places. Only the abnormal conduction needs to be stopped. Mechanical forces passing through these paths can be retained. How about using some drug to permanently block electrical conduction but keep the mechanical properties, say, a permanent ion channel blocker? Short term channel blockers are used widely for many applications. Long term blockers are not so common, though some studies do report the usage of long term blockers for pain management.

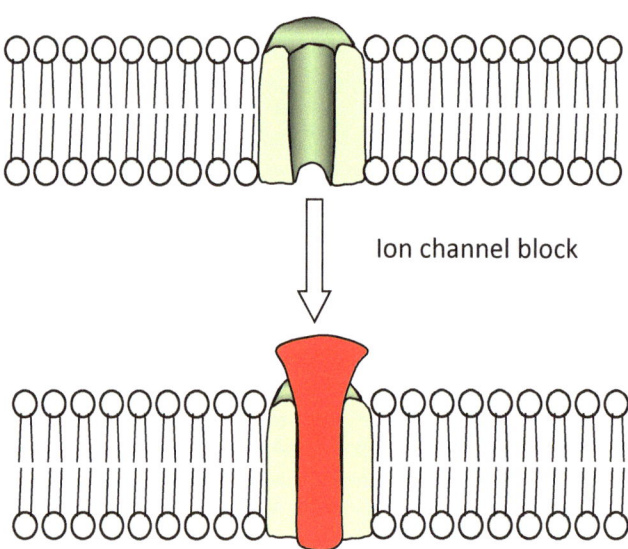

Ion channel block

Rejuvenation after ablation

After ablating wrongly conducting heart tissue, that region is currently left to heal on its own. How about injecting or applying stem cells to restore the mechanical function of that region? The new cells may be chosen to increase electrical resistance, have no native pacing ability and block conduction. Possibly by gene knock out of ion channel proteins responsible for transmitting action potential. The best case replacement, though, should aim to retain the action potential induced contraction and remove only gap junctions responsible for action potential transmission or action potential initiation. This guarantees maximum force production with minimum arrhythmia risk. Some prior work seems to have attempted recreating tissue resembling pacing nodes using genetic techniques. Creating heart cells with blocked conduction seems to be unexplored.

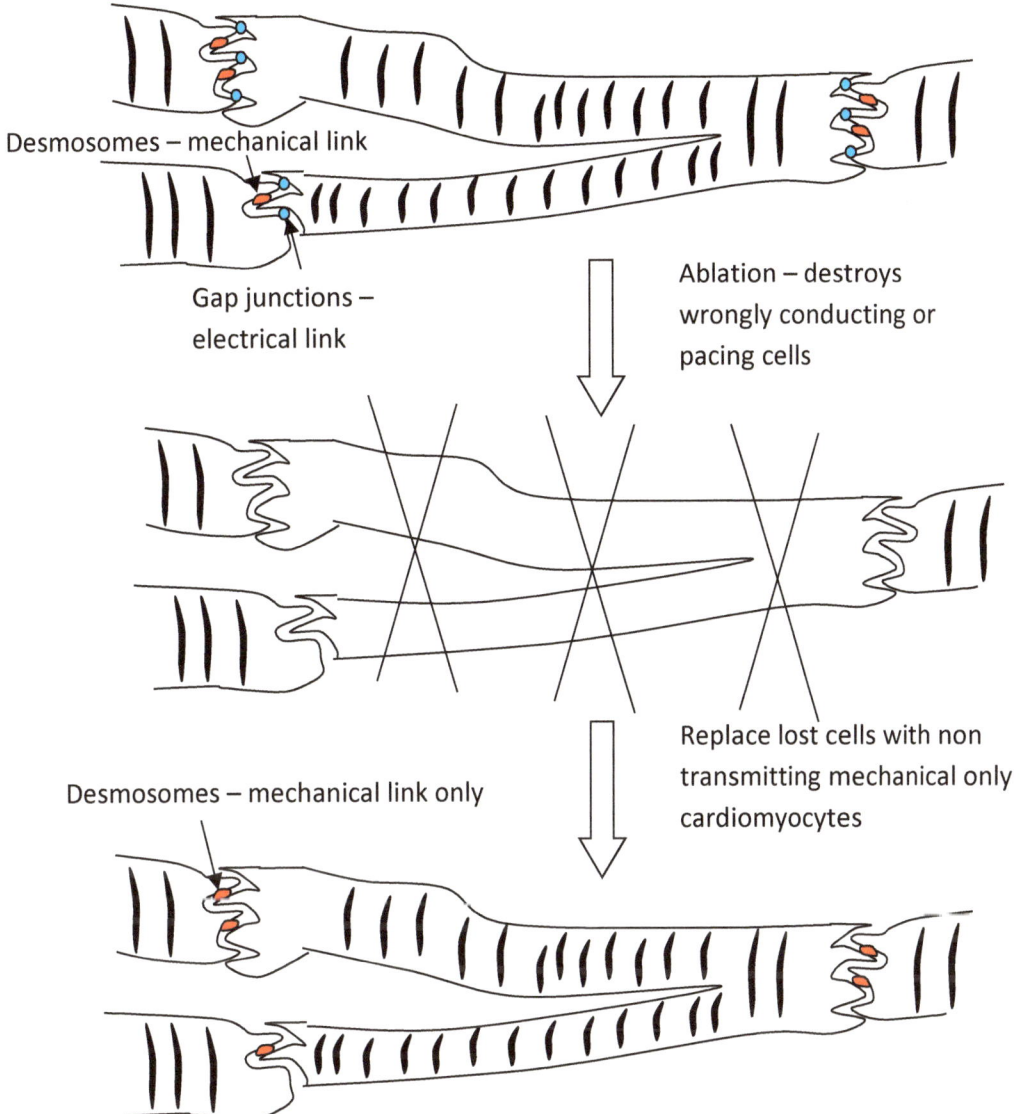

Desmosomes – mechanical link

Gap junctions – electrical link

Ablation – destroys wrongly conducting or pacing cells

Replace lost cells with non transmitting mechanical only cardiomyocytes

Desmosomes – mechanical link only

Re-establishing cardiac conduction with wires or nerves

When there is conduction block that causes abnormally low heart rate artificial pacemaker is used to provide the required pacing signals to the myocardium. How about using wires to bypass conduction blocks? How about using nerve tissue made from cells to replace blocked paths? This may be a cheaper and less invasive alternative than pacemakers because the wire or neural prosthesis for blocks alters the native heart conduction system much less than artificial pacemakers.

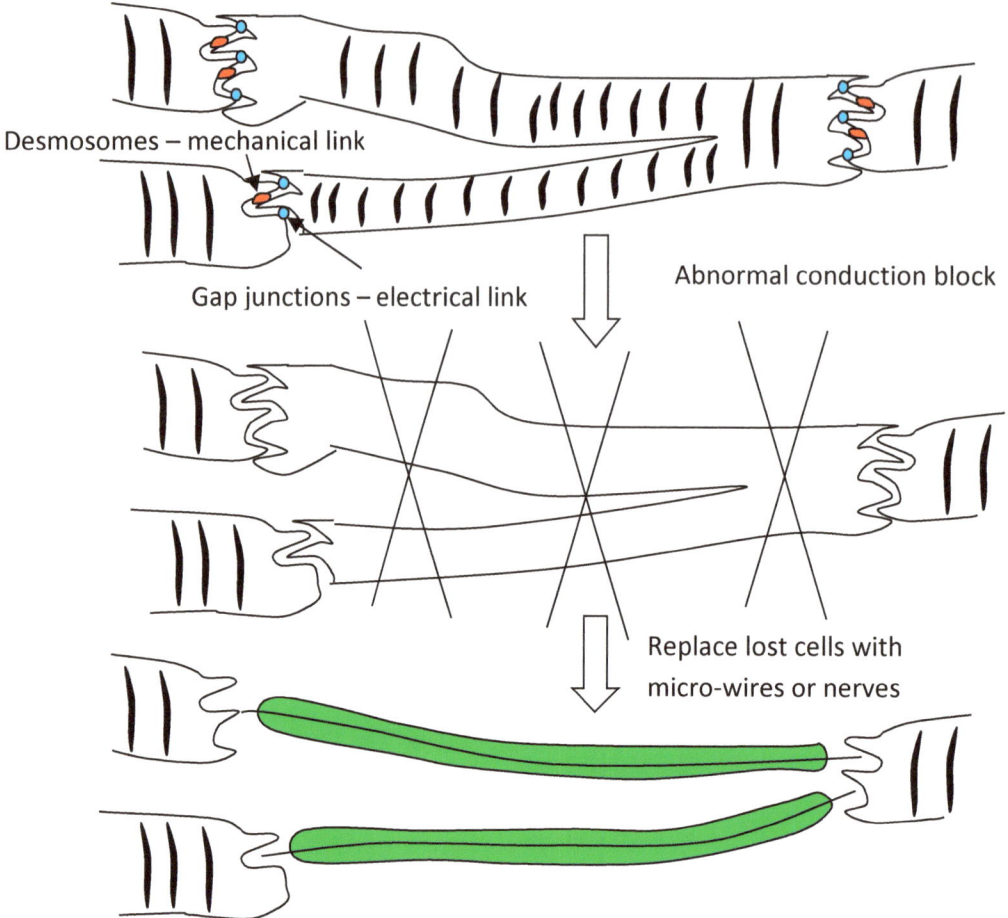

Desmosomes – mechanical link

Gap junctions – electrical link

Abnormal conduction block

Replace lost cells with micro-wires or nerves

Cell coated contacts for ICD/IPG

ICDs and IPGs have an electronic to biological interface, i.e. the internals of an ICD/IPG is made of electronic components but the signal needs to finally reach the heart muscle. The predominant approach is to use biocompatible metallic contacts that hook on to the heart muscle. How about using an alternative contact that has a special texture or shape on which neurons or cardiomyocytes are coated and then this composite contact is hooked onto the heart muscle? This may lower contact resistance and hence lower power requirement to provide the pacing signals. It may also be possible to increase the electrical conductivity of the coated cells by using genetically modified cells with large number of ion channels.

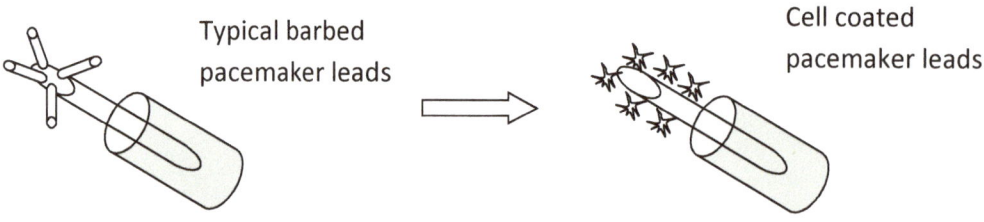

Typical barbed pacemaker leads

Cell coated pacemaker leads

Part 3 - Gasping for blood or drowning in blood – ischemia and reperfusion injury

Reperfusion injury occurs because blood flow to ischemic heart tissue is restored fast. Can reperfusion be done slowly? This section deals with possible ideas to avoid reperfusion injury.

Pre-blocked bypass vessels

In the regular bypass surgery, the surgeon grafts a vein as a bypass for the blocked coronary artery. If we pre block the graft the bypass will not reperfuse the ischemic tissue. But we do want reperfusion. Reperfusion injury can be avoided by designing the blockage in the graft to open slowly. The pre-blockage may be achieved by a constricting ring on the bypass graft vein with constricting ring designed to slowly lose tension and dissolve ultimately. The constriction can be inside to bypass vessel too. Another method could be to use a bio compatible substance that dissolves slowly. Some of the biocompatible substances suitable for pre blocking are non-toxic gels with slow water solubility.

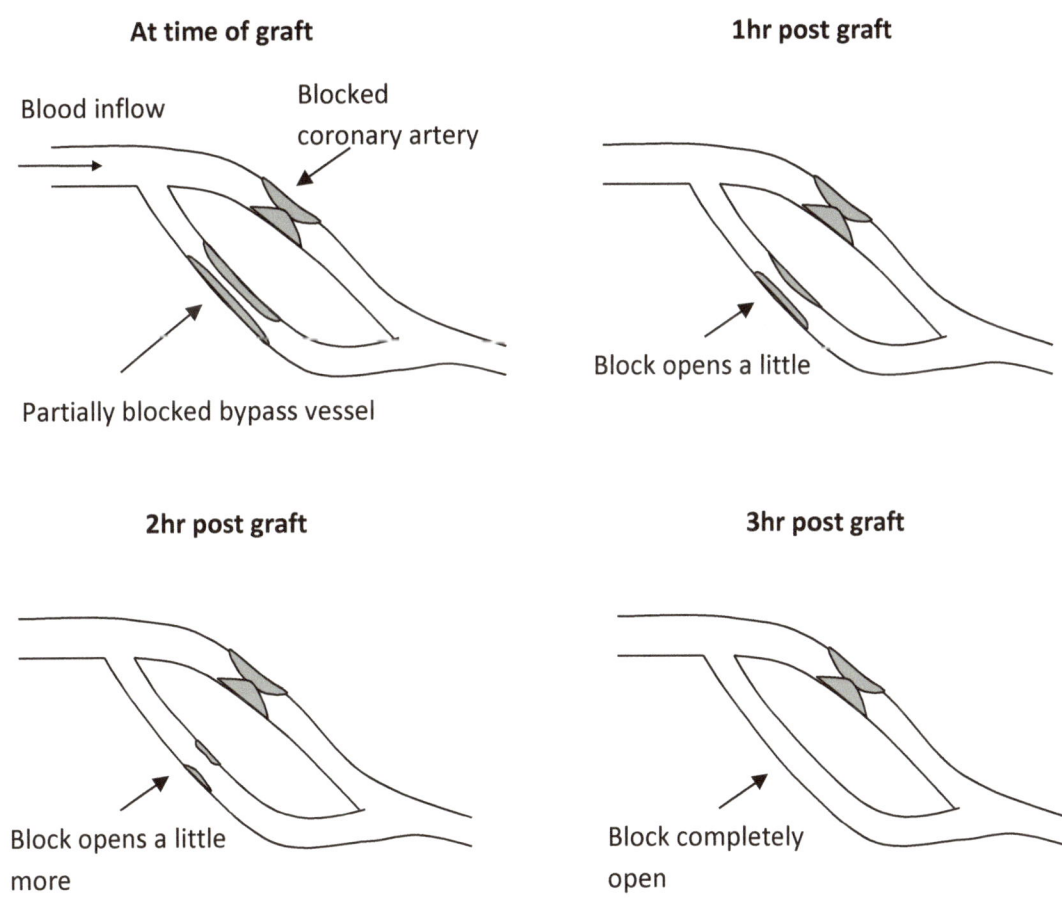

Oxygen carrier modulation

Some of the synthetic blood substitutes are known to be smaller than RBCs. They can diffuse a through tissue better and some of them can carry more oxygen than RBCs. Many of the substitutes though are not ready for long term use but may be ok for short term in-surgery use. These can be tried to reduce reperfusion injury. Sometime before the bypass surgery, the patient can be infused with high oxygen capacity and small blood substitute molecules. This can be expected to improve the oxygenation of the ischemic tissue. Immediately after the bypass starts carrying blood, infuse drugs to block the blood substitute molecules that were used before. The overall effect may be to reduce the speed of increase in oxygen concentration. Same technique can be used for other nutrients too.

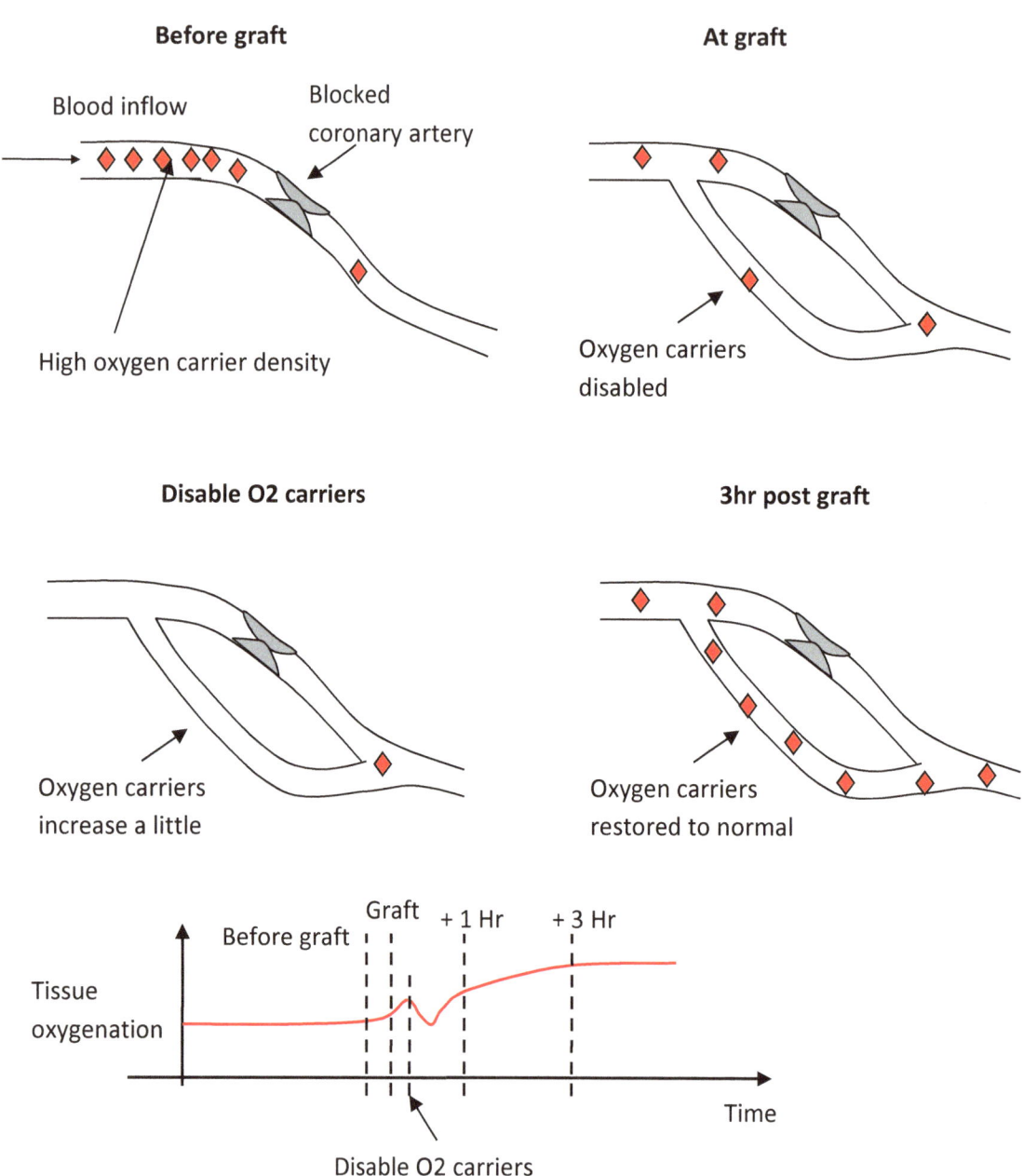

Tapering concentration cardioplegia

Cardioplegia is typically done in three phases, induction, maintenance and reperfusion. The induction phase requires quickly establishing equilibrium of all the necessary molecules for tissue preservation. In the conventional approach, cardioplegia is started with one solution and then maintained with another solution. To achieve quick equilibrium the induction solution is pumped at higher pressure than the maintenance solution. But there is a limit on how much flow that the vessels can carry without being injured. To accommodate both the need for higher amount of molecules while keeping the pressure within limits, the only option left is to increase the concentration of the solution. How about using only one solution that starts with higher concentration and then tapers to lower levels?

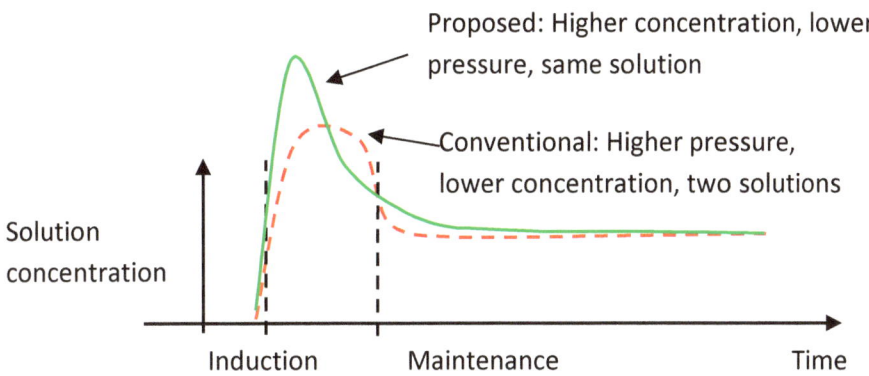

Regenerated cardioplegia solution

For prolonged surgeries, the cardioplegic solution may degrade overtime because of being used by the myocardium. Acidosis is one such phenomenon. This can be expected because there is limited homeostatic control of the concentration of ingredients. Oxygen is regulated through the use of heart-lung machine in the cardioplegia circuit but the homeostasis of other chemicals that is normally ensured by kidney and liver is absent. It will be interesting to see if cardioplegia solution can be regenerated with an external module with liver and kidney cells that forms part of the blood flow loop. Some kind of dialysis machine may be a good starting point. After oxygen, the next important ingredients are simple sugars and fatty acids that are metabolized by heart tissue. This can be added at a constant rate to the cardioplegia circuit. The waste products of these being only carbon dioxide get removed by the oxygenator. To get the behavior of liver and kidneys, engineered liver and kidney tissue may be useful. Since these tissues need to work only during the surgical procedure, they could be made from existing embryonic stem cell lines without having to worry about immune compatibility.

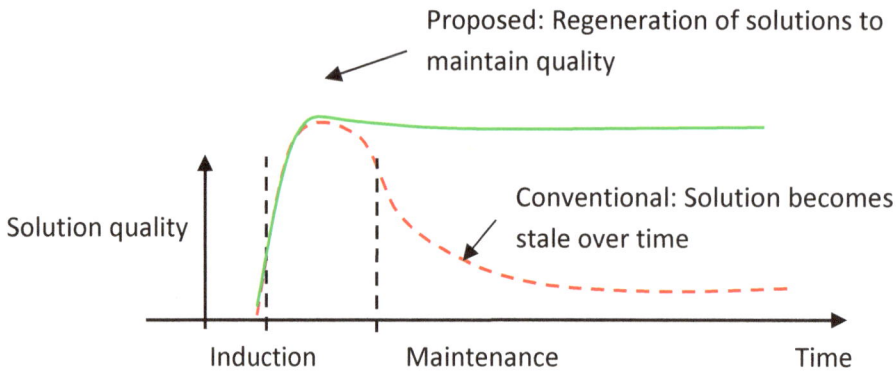

Proposed: Regeneration of solutions to maintain quality

Solution quality

Conventional: Solution becomes stale over time

Induction   Maintenance   Time

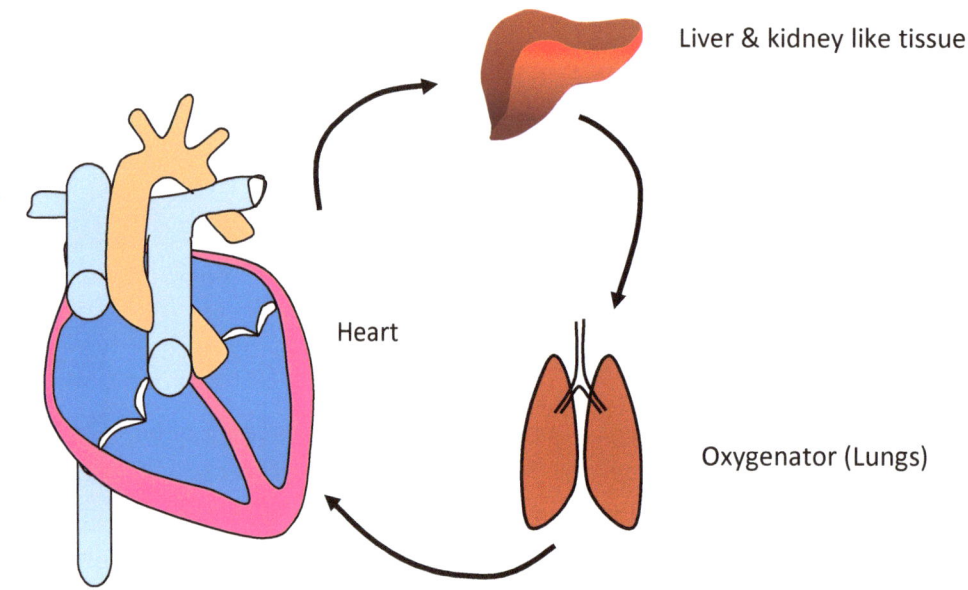

Liver & kidney like tissue

Heart

Oxygenator (Lungs)

Chemical tweaks for cardioplegia

Cardiac switch - molecules to arrest heart

Apparently, cardiac arrest is mostly induced by elevated potassium ion concentration. Though this is reversible and only mildly toxic, it is still different from the normal physiology that cardiomyocytes are used to. The most minimal modification to the heart will be some kind of molecule that selectively turns OFF cell contraction mechanism without altering other cell behavior. It should also be reversible, i.e., should turn ON contraction after the surgery is complete. Ion channels or internal proteins part of the contraction cascade are prime targets for this. Many toxins in the natural world disrupt the ion channels. Auto antibodies have also been shown to affect ion channels in some diseases. Understanding these mechanisms may allow for creating a designer cardiac switch drug with low side effects. This may be useful in keeping the viability of myocardium throughout the surgery and also have lesser side effects like brain damage.

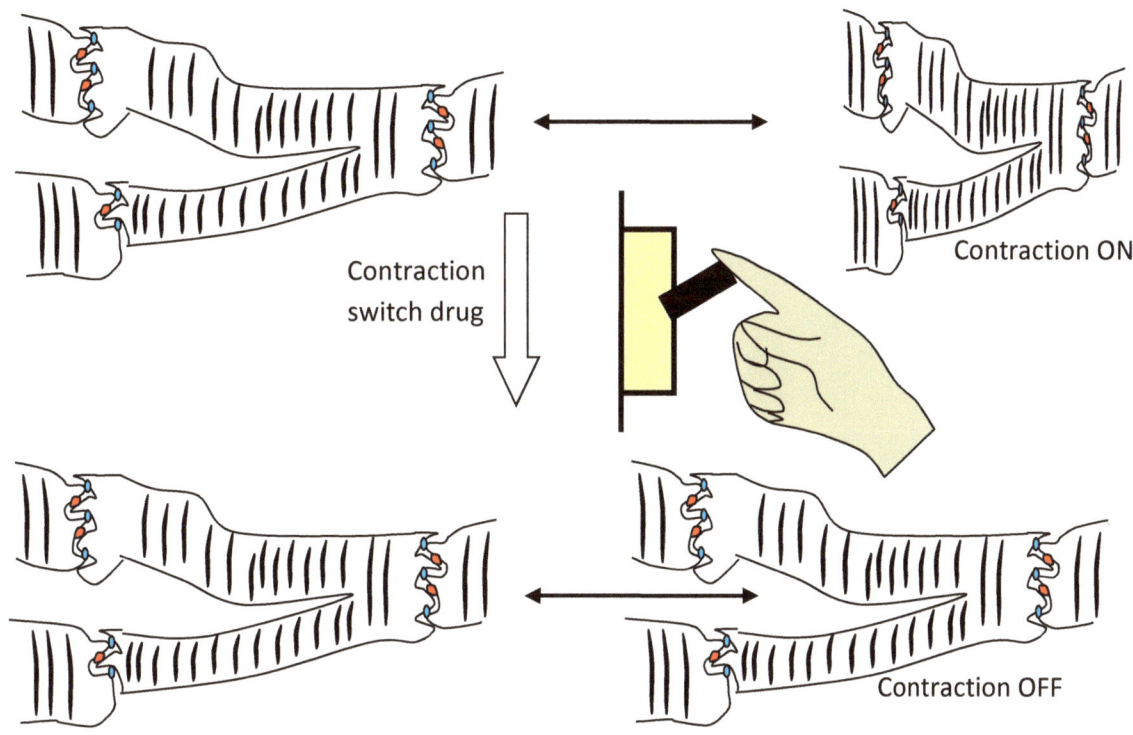

Molecules to preserve cardiomyocytes

It is well known that hypothermia preserves cardiomyocytes when the heart is stopped. But, it is a rather crude way with some side effects. When the heart is stopped tissue damage happens in specific ways. Some of these ways are cell autophagy, apoptosis and necrosis. Autophagy is a cell's survival mechanism when faced with nutrient shortage. It eats its own contents to survive just a bit longer. But this is maladaptive during cardioplegia. Apoptosis is programmed cell death that is meant to kill cells that have gone rogue. Again, this is also maladaptive during cardioplegia. Precise blocking of these pathways during cardioplegia may provide a side effect free cardio protection. Some of the cardioprotective molecules can be antioxidants, anti-apoptotic drugs, anti-necrotic drugs and anti-inflamatory drugs. Some of these have been identified as Resveratrol, Cyclosporin, Necrostatin-1, Infliximab and more. In a completely alternative way, there could be some cocktail of chemicals that stops cardiomyocyte metabolism completely without killing the cells. Suspended animation is used by many organisms to deal with nutrient deficiency. Hydrogen sulfide has been shown to reduce cellular metabolism drastically without side effects in some lower organisms. H2S has even been shown to be beneficial during cardioplegia.

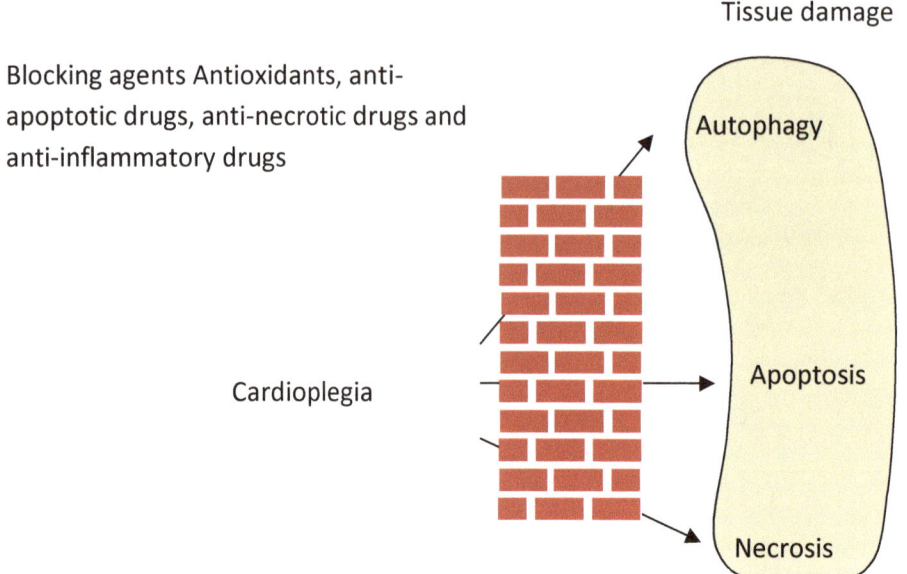

Tissue damage

Blocking agents Antioxidants, anti-apoptotic drugs, anti-necrotic drugs and anti-inflammatory drugs

Autophagy

Cardioplegia

Apoptosis

Necrosis

Energy molecules

Though oxygen and to a lesser extent glucose are the main limiting nutrients during ischemia, energy molecules are the ones that are really needed by the cells. From this section onwards, energy molecules will refer to molecules like ATP, NADPH, FAD, CP, GTP that power the biochemistry of cells. So, in principle, supplementing the cardioplegia solutions with energy molecules may improve cardioprotection. ATP seems to have been used as a therapy for some pre-terminal patients but, ATP seems to promote aggregation of platelets as well. Creatine phosphate seems to have benefit as a cardioplegia solution ingredient. NAD has some therapeutic effects reported for Parkinson's disease. Still, there are only a small number of reports about the use of energy molecules used for cardioplegia. This may be a good direction to explore. Energy molecules may also be a good candidate for intravenous infusion during cardio pulmonary resuscitation and for critically ill patients.

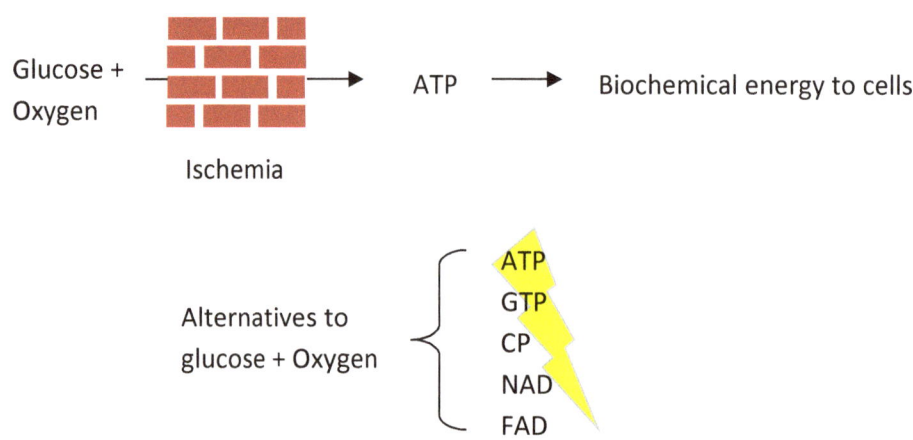

Glucose + Oxygen → ATP → Biochemical energy to cells

Ischemia

Alternatives to glucose + Oxygen
{ ATP
GTP
CP
NAD
FAD

Part 4 - Rethinking energy delivery system

Heart-lung-mitochondria system

The cells of the body need a continuous supply of material. Some of the materials contain energy and others are predominantly used as building blocks. An even smaller fraction is used for communication. The heart can be thought of as powering the material transport network, the lungs as a fuel import-export port and the mitochondria as a local engine in every cell that converts the bulk fuel into a form of energy that is convenient to power the cell biochemistry. If understood this way the whole system can be repartitioned in various other ways.

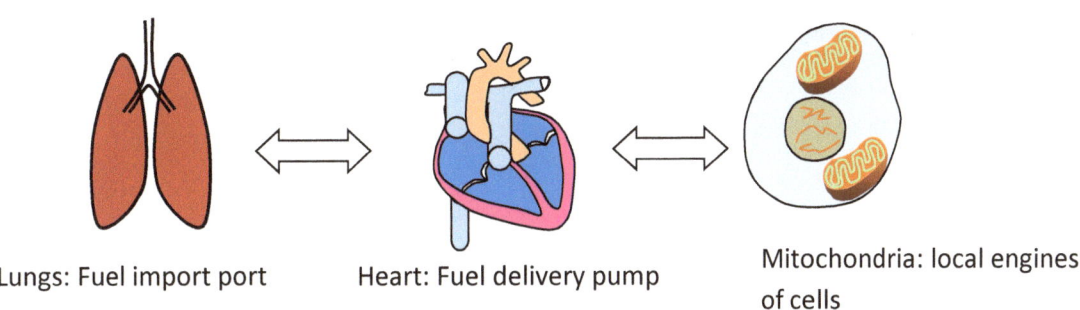

Lungs: Fuel import port   Heart: Fuel delivery pump   Mitochondria: local engines of cells

Combo energy-pump

We can envisage a single organ that inhales-exhales gases, converts glucose to convenient biochemical energy molecules and finally pumps out these energy molecules to every part of the body. More specifically, it can be thought of as lungs with pumping muscles and cells with large number of mitochondria for energy molecule production. This is like lumping all the mitochondria of the body into one organ.

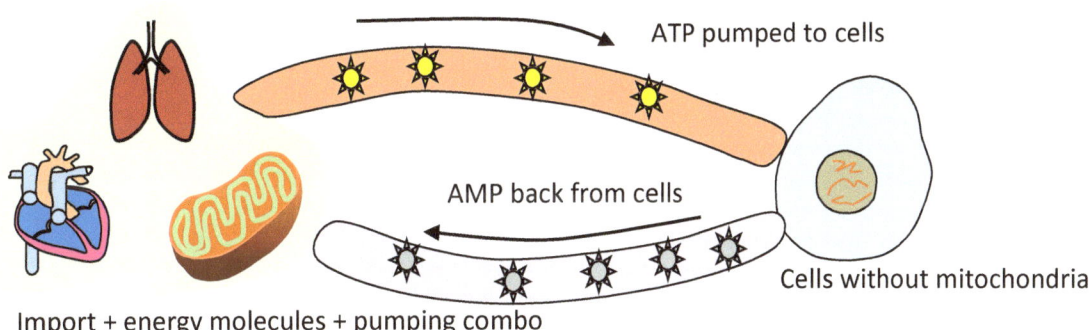

ATP pumped to cells

AMP back from cells

Cells without mitochondria

Import + energy molecules + pumping combo

Megachondria

Keeping the function of the heart and lungs almost the same, we can add a new organ, say we name it megachondria. The megachondria's job will be to do the work of the mitochondria of all the cells and create energy molecules. To reduce the complexity in circulation, megachondria is better realized in the regular lungs.

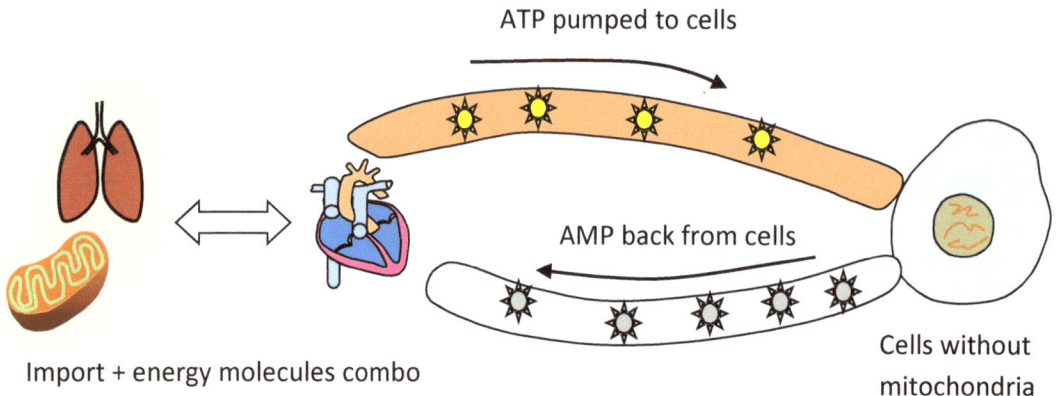

ATP pumped to cells

AMP back from cells

Import + energy molecules combo

Cells without mitochondria

Lart

Keeping the mitochondria to be distributed in every cell as usual, we can envision a Heart + lung combo organ that both exchanges gases and also pumps blood. It may be easier to engineer because the two conjoined pumps of the natural heart is not needed anymore. Or it may need lesser pressure and be more resilient to failure.

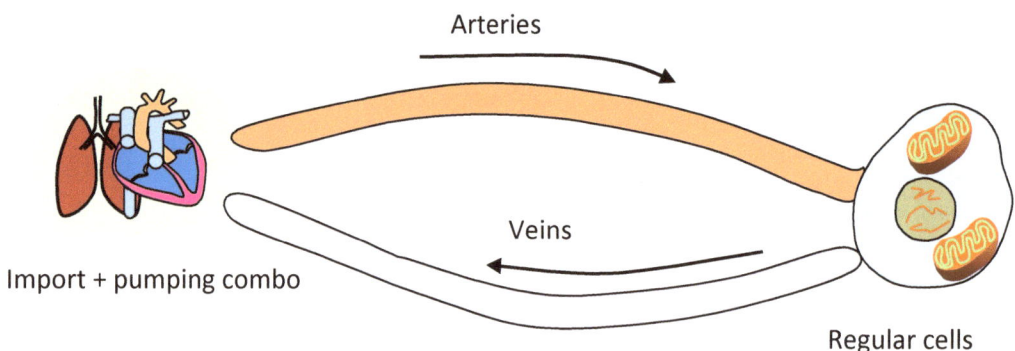

Arteries

Veins

Import + pumping combo

Regular cells

Alternative currency for biochemical energy

ATP is the dominant energy molecule in biology. But, its energy density is far less than other sources like hydrocarbons. The advantage of ATP is that it can readily transfer power to other reactions. One of the reasons for this is that the phosphate group has energy stored in its bonds. Is it possible to use other molecules that are denser and as much convenient for energy transfer? For example the ease of energy release of high explosives is primarily due to the nitro, peroxide, nitrate, chlorate, perchlorate, azide groups. Transition metal ions with their broad range of oxidation states provide another convenient source. Manganese, cobalt, iron oxidation reduction reactions are frequently used in batteries.

$\Rightarrow$

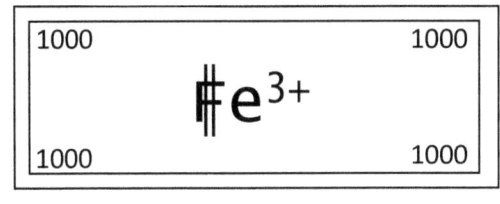

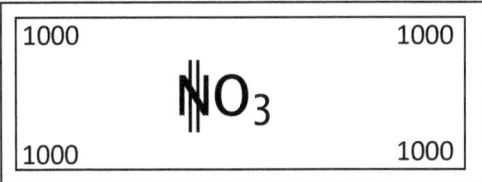

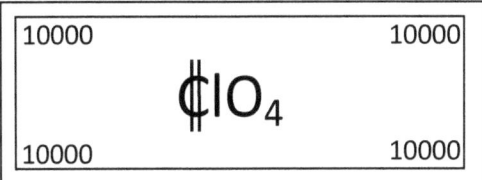

Part 5 – Genetic and cellular tweaks

This part discusses the tweaks that can be done to the genes and cellular machinery with the aim to make the cell more efficient. Since it may not be feasible to administer a large number of cells to patients, improving the efficiency of every cell makes the therapy more effective.

Recursive automated DNA removal

It is speculated that large parts of eukaryotic genome is junk. What if we remove the unwanted parts automatically? For example, starting with yeast cells a machine can be made to put aside half of the yeast cells and in the other half of the cells strip just a few segments of DNA from its genome and check for growth and reproduction through a few generations. If the removal worked it could discard the half that was unchanged and double the cell count of the reduced genome population. The doubling is to ensure same count of cells in every step of the algorithm. If the removal did not work and reduced genome cells are not viable the machine could backtrack to the cell population with unmodified genome of the previous step and discard the unviable cell population. After backtracking to previous known viable population, the machine could try next step by removing some other piece of the genome that was not removed before. Repeating this process till the time that the machine is unable to find any subsequent viable genome after, say, 20 trials can be called the most reduced genome of the cell. To speed up the removal process, many parallel experiments can be launched at the same time with each experiment removing a piece different from the other experiments. Along the way, the machine can record the genomes and the phenotypic behavior of the corresponding cells. The most reduced genome cells could be more energy efficient than the cells with junk DNA because they do not have to copy the junk parts during every cell division. By analyzing the large amount of genotype-phenotype data produced it may be possible to automatically elucidate gene function and genetic networks in the cell. Other use is the removal of endogenous deactivated viruses. Cancer cell lines, embryonic stem cell lines, yeast are good candidates for these experiments because they have the capacity to divide indefinitely unlike other cells of adult organisms. Here junk is used as term to mean practically little use, i.e., the organism will be almost unaffected if it is removed. This definition avoids the issue of coding/non-coding and beliefs of function. The definition also considers marginal benefit as no benefit so that we can focus on reaching the most reduced genome with almost the same behavior as the original organism. Without going into the debate of junk DNA or not, automatic recursive knockout will be a useful tool to study gene function.

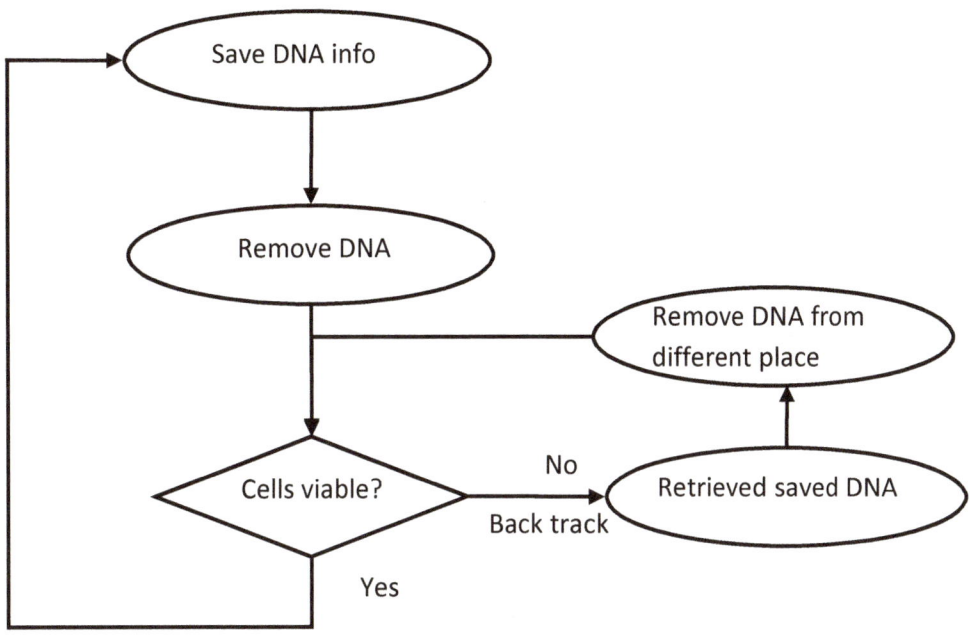

Selfied cells

Although the perception of induced pluripotent stem cells being slightly inferior to embryonic stem cells is waning, it should be noted that autologous embryonic stem cells are considered the best stem cells possible. However, creating autologous ESC involves therapeutic cloning that is fraught with ethical and legal dilemmas. Fortunately, some ESC lines are available and cloning is no longer required. But the autologous part is the missing piece in the puzzle. The need for autologous cells is that HLA markers are specific to every person coded in MHC genes. What if we are able to transplant the patients MHC genes into the readily available allogenic ESC? We may get stem cells that are as good as autologous ESC from immune compatibility point of view and without having to clone. Transplanting all the MHC genes of the host into the allogenic ESC may be hard because MHC genes are numerous. Instead, it may be easier to completely replace chromosome 6 that carries all the MHC genes. One approach can be to mechanically isolate C6 from host, remove C6 from allogenic ESC then mechanically transplant host C6 to allogenic ESC. The host C6 can be taken from an adult cell. For ease of identification, the host cell can be made to mitotically divide and when the chromosomes bundle and take their characteristic shapes arrest the division and mechanically remove the C6. Same processing can be done on allogenic ESC. Finally, host C6 can be transplanted to allogenic ESC and the mitotic arrest lifted to let the ESC continue dividing. After a few cell divisions, the selfied ESC may completely lose the allogenic HLA markers and show only the host HLA markers. Some biochemical methods can be used to supplement the C6 removal. For example special proteins may be designed to bind to specific sites of C6 that are constant throughout the population. The C6-protein complex can be manipulated easily if the protein was endowed with some handles, i.e., chemical functional groups. To improve the quality of the selfied ESC, the host C6 can even be started from the C6 of a host iPSC. In this way, even the epigenetic state of C6 can be closely matched with that of the true ESC C6. It is worth noting that other simpler alternatives are being explored to avoid the immune compatibility problem by having the immunogenic MHC genes being knocked out of allogenic ESC. This may have the disadvantage of reducing the innate immune system capacity to create novel antibodies to defend against pathogens. Hence, creating selfied ESCs as close as possible to true autologous ESCs need to be pursued. In a typical scenario for the usage of selfied ESCs, a cell bank can create selfied ESCs and keep them ready for every paying customer. Whenever the customer needs some stem cell based therapies, these cells can be used. An advantage of this technique is that we can take the best performing cells for a task and personalize it to the individual, i.e., for hematopoietic stem cells we can use the cells with the best performing hemoglobin genes and the capacity to produce good quality RBC, WBC and platelets.

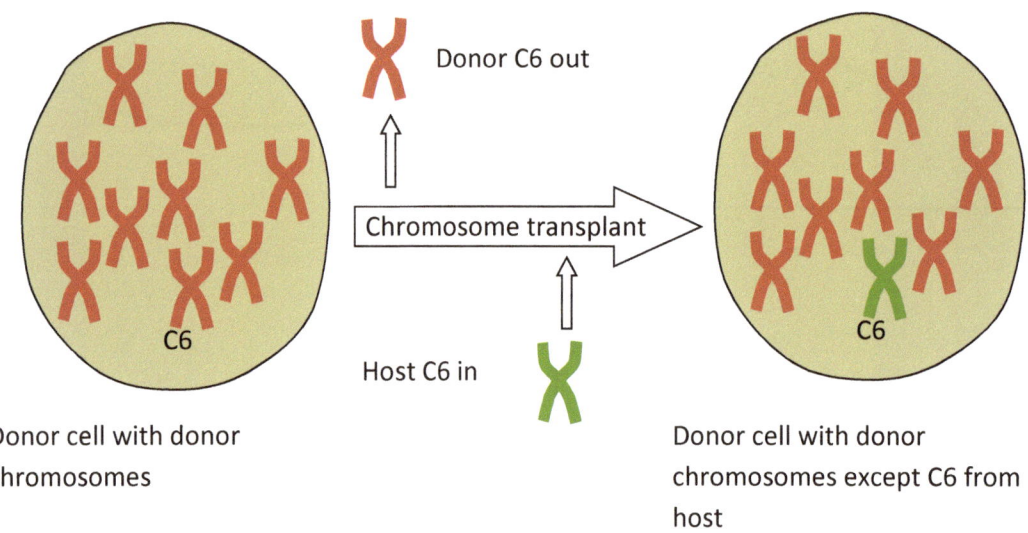

Donor cell with donor chromosomes

Donor cell with donor chromosomes except C6 from host

Candidates for genetic tweaking

Longevity

It has been shown that modifications to the expression of specific genes, like DAF-2, have profound impact in the longevity of some organisms. The changes in the genes of these organisms are small in number and so applicable to therapies.

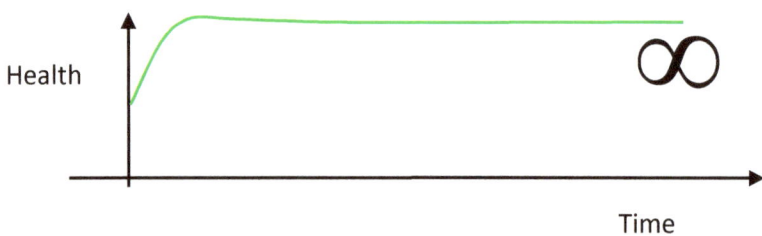

Apoptosis

Optimizing apoptosis related genes can improve the safety of cell therapies by reducing the possibility of tumors.

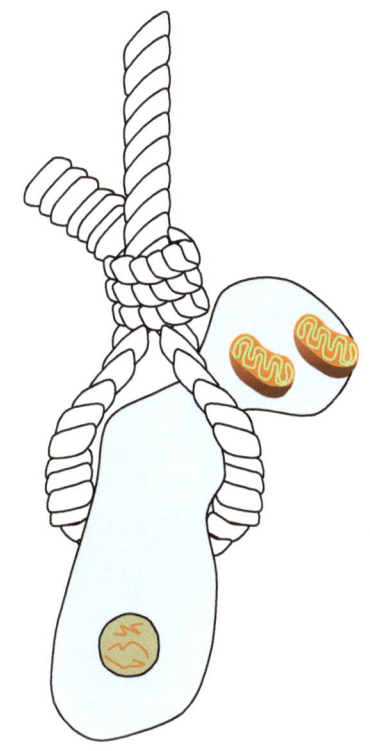

Motility

Cells like that of the immune system need motility but many other tissue types need cells that are stuck in their locations. At most, some of them may need to grow into nearby tissue to replace dead cells. Knocking out most of the motility related genes from cardiomyocyte stem cells may help reduce the incidence of metastatic tumors originating from therapeutic cells.

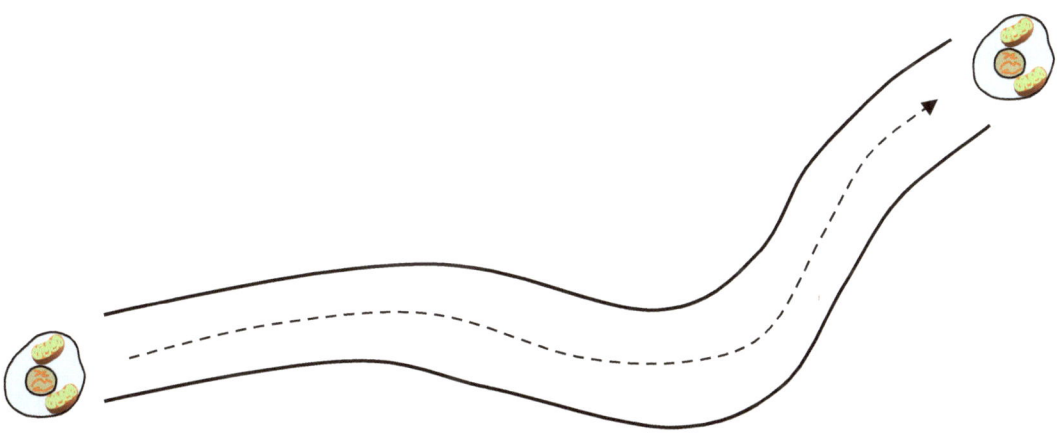

Mitochondrial biogenesis

The number of mitochondria in a cell is dynamic. Aging is associated with decreased number of mitochondria in cells. It may be possible to enhance the genes associated with mitochondrial biogenesis to reduce the effect of aging on therapeutic cells.

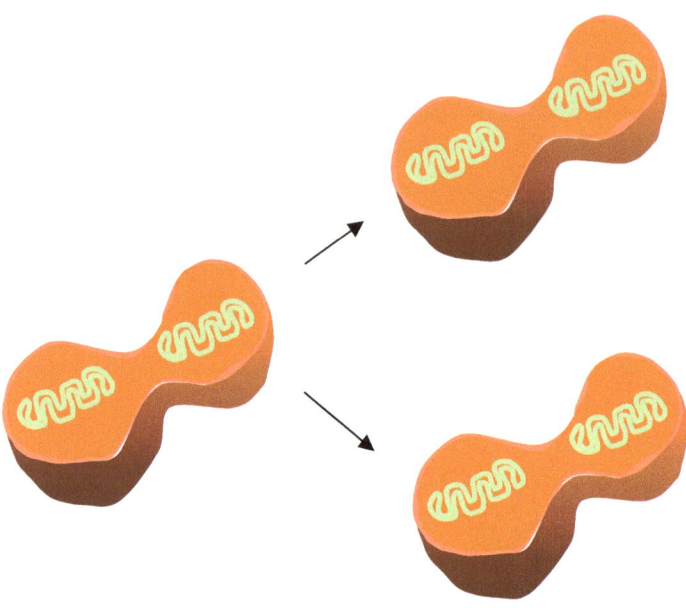

Mitochondrial genes copied for redundancy

Mutations in the mitochondrial DNA have been implicated in the aging process. Since mtDNA is not diploid like the nuclear DNA there is no backup copy to rely on. In nature, the backup mechanism is actually by having a large number of mitochondria in the oocyte to begin with. The mtDNA seems to code for 13 proteins. It may be interesting to see if creating mitochondria with 2 or more copies of the 13 genes improves the longevity of the cell lines. If this is indeed true, mtDNA with best function and good match to the host genome can be sequenced and then synthetically prepared or isolated from natural mitochondria and then incorporated into the selfied embryonic stem cell. The resulting cell may have enhanced vigor.

Many groups are focused on improving health by changing genes. In particular, SENS foundation's proposals regarding genetic manipulation of cells are directly applicable to selfied ESCs.

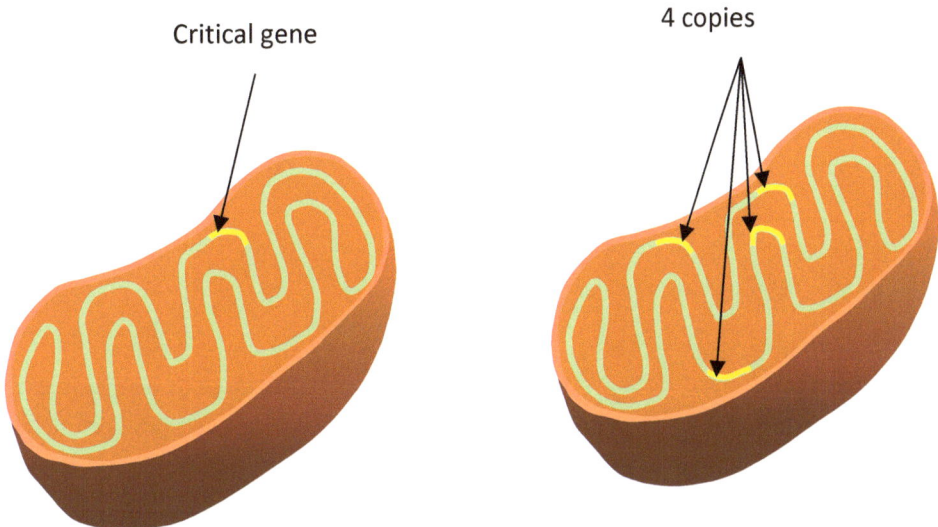

Critical gene

4 copies

Genetically enhanced oxygen carrier protein

Oxygen transport in blood happens via hemoglobin. Other proteins like neuroglobin, cytoglobin, myoglobin and hemocyanin transport oxygen as well. Natural variations in oxygen binding properties of hemoglobin do occur even amongst humans, with higher affinity corresponding to races of higher altitudes. This hints that hemoglobin based oxygen transport of the general population need not be the optimum achievable. It may be possible to genetically engineer the oxygen carrier protein of RBC to increase oxygen carrying capacity. The size of the molecule can be made smaller so that more oxygen can be packed. Alternative shapes can be tried to reduce the size of the cell that reduces susceptibility to clots. Possible targets are replacing the porphyrin ring with some other coordination friendly molecule like EDTA which is one of the most potent ligands, replacing iron with some other transition metal like manganese, copper or zinc. It may also be possible to remove metal from hemoglobin altogether leaving only a pure protein oxygen carrier. For example, there are cofactor independent Oxygenases suggesting that a transition metal is not mandatory for oxygen binding. This may make its metabolism much easier. During some pathologies like muscle damage or RBC damage, the hemoglobin or myoglobin pouring into the blood stream is toxic to the kidneys mainly because of the iron content. Remove the metal remove the toxicity. Another approach could be to have the lung tissue convert dioxygen of air using oxidoreductase enzymes into hydrogen peroxide that is later transported through some other carrier. Dioxygenase class of enzymes could be tweaked to reversibly bind oxygen for transport instead of its normal function of catalyzing the oxidation of some substrate.

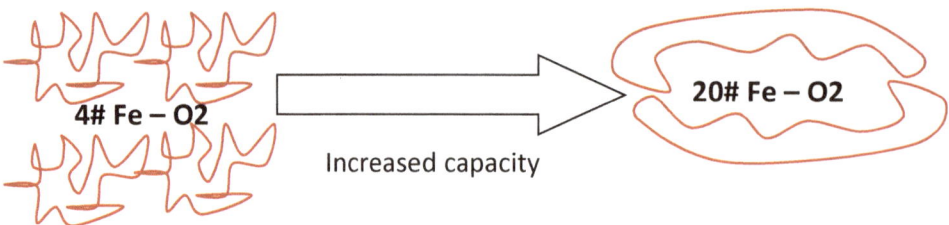

Genetically enhanced RBC

RBCs vary quite a lot in the animal kingdom. The human RBC ejects its nucleus and hence cannot survive for long. Many other animals do have RBCs with nucleus. Is it possible through genetic engineering to make RBCs that keep the nucleus with the intention of extending the lifespan of the RBCs? This will reduce the need for high cell division rate in hematopoietic stem cells. The shape of the RBCs differ widely too. It is worth exploring if some other animal RBC is better in terms of oxygen delivery and better cardiovascular function and resistance to parasites like that which causes malaria.

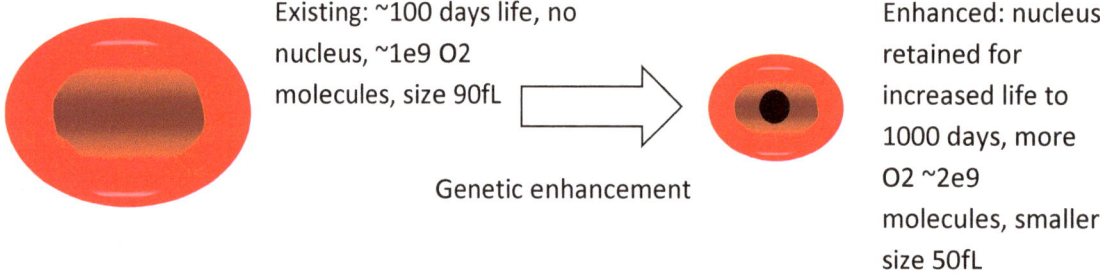

Cardiomyocyte performance across species

Cardiomyocyte performance varies across species. The ion channel currents, amount of reserve ions, size etc. vary. By comparing various performance metrics of cardiomyocytes like contractile force, speed, energy efficiency, durability, resistance to ischemia and resistance to infection across species, it may be possible to incorporate these findings in tissue engineered myocardium. Typical candidates for study are the extreme adaptation animals – elephant cells or blue whale cells for they are part of the largest hearts, cheetah/humming bird/sail fish for possible high heart rates, tortoise for durability and Cuvier's beaked whale for record breath holding time.

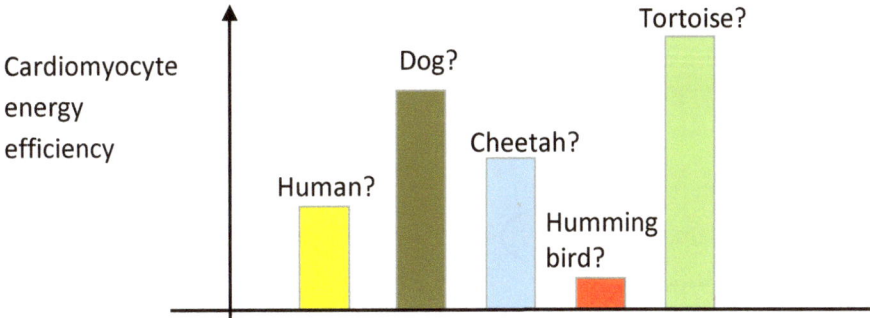

? – Purely for illustrative purpose. Actual values may be different than illustrated here

Iterative creation of complex vasculature

Creating fine capillaries is one of the main challenges in bioengineering complex organs. One approach that has seen some success is taking existing organs from animal or cadaver and then stripping its cells to get a scaffold that has the fine structure of capillaries. Another approach being developed is 3D printing of these scaffolds or full blown tissues. This approach though promising has not yet reached the quality of natural tissue. An iterative approach that starts off with 3D printing and then iteratively refines the scaffold may reach the quality of natural tissue. Iterative approaches have been used in modeling angiogenesis because iterative mechanism is natural and also readily implementable in computer simulations. Iterative methods to bioengineer ECM seem to be uncommon, probably, because the cost of the final ECM is proportional to the number of iterations. Iterative approach is scalable in mass producing high quality ECM using existing human embryonic stem cell lines. If the starting stem cell lines are pure the end product ECM carries lesser risk of infection compared to cadaver or animal derived ECM.

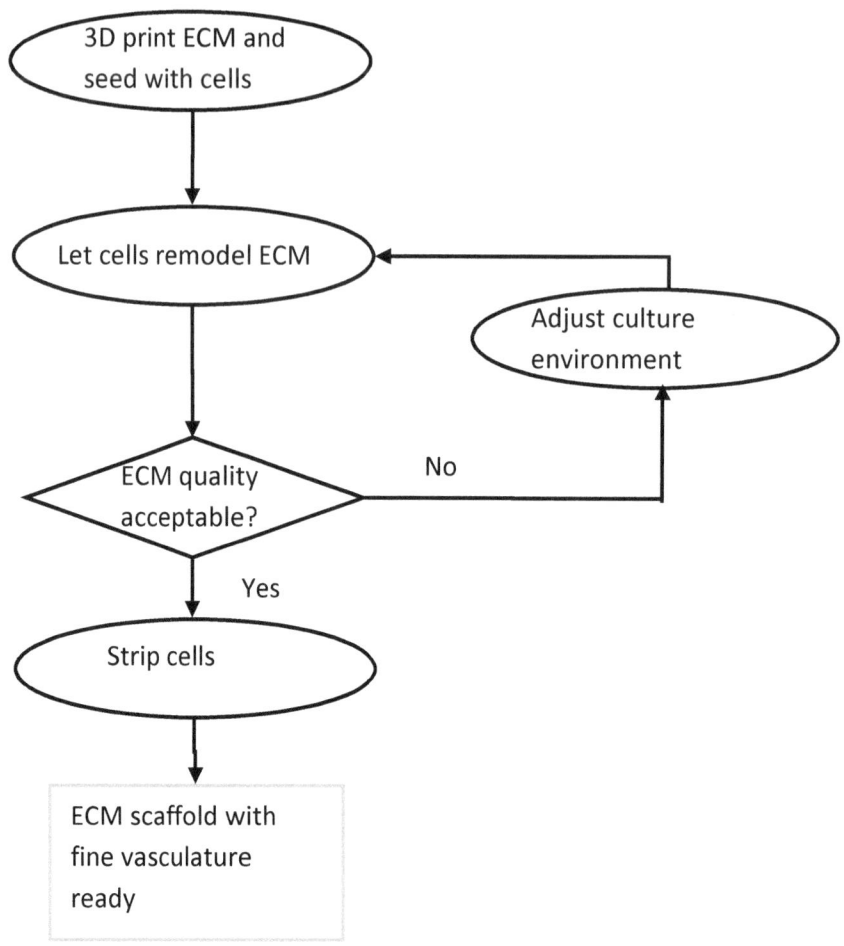

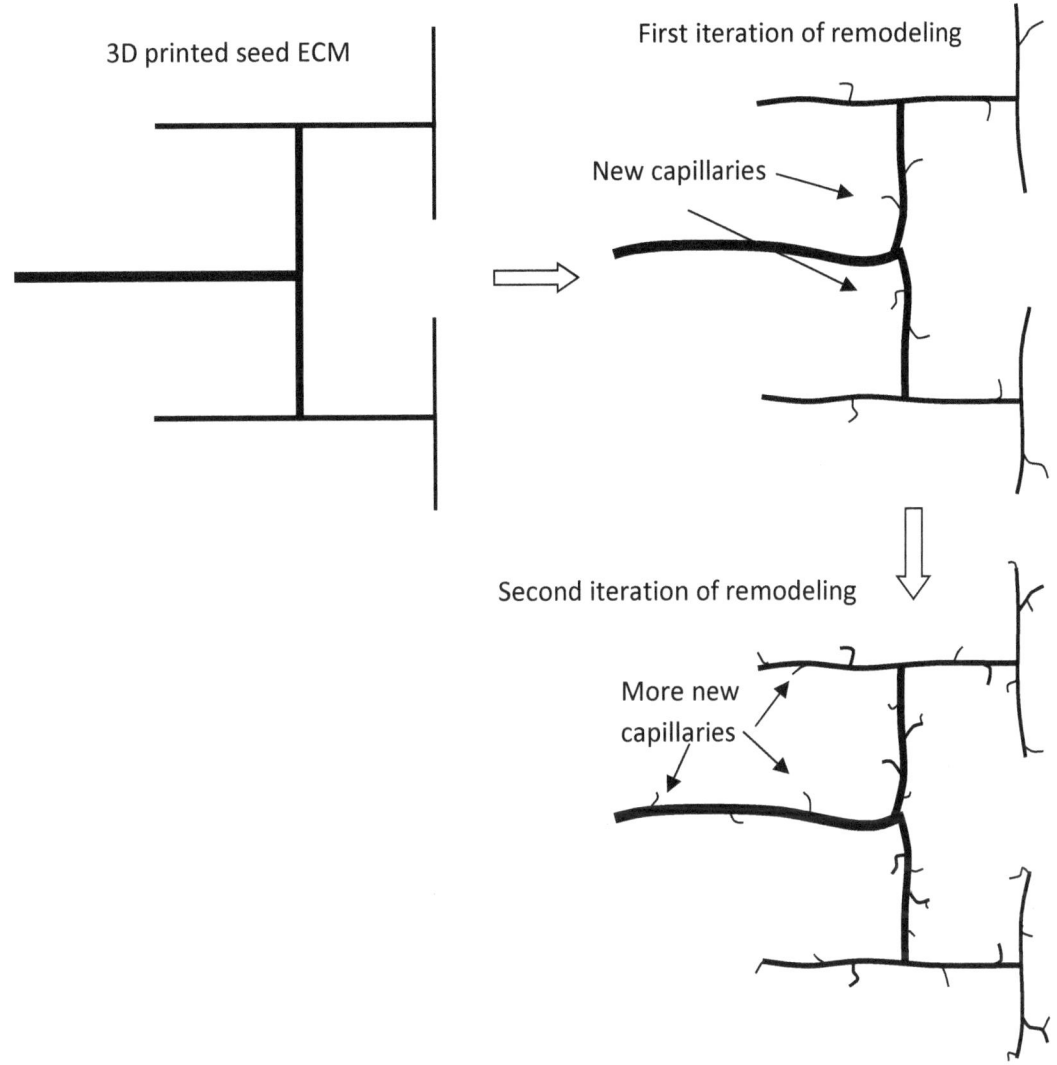

3D printed seed ECM

First iteration of remodeling

New capillaries

Second iteration of remodeling

More new capillaries

Part 6 – Optimizing circulation

Collateralized bioengineered heart

Collateral circulation in the heart refers to blood supply to the heart muscle via more than one blood vessel. Collateral circulation confers some degree of resistance to muscle death when a coronary artery is blocked because there are redundant paths for the blood to reach the muscle. The variation of collateral circulation across animal kingdom is high. Guinea pigs have one of the highest collateral circulations. How can we exploit this to reduce vulnerability to heart attacks in humans? For example, the guinea pig embryonic development of the heart can be studied to understand how these collateral vessels get created. There may be particular growth factors that may be useful for human use. How can the genetic developmental mechanisms involved be used in humans to reduce the incidence of myocardial infarctions? Some high altitude animals have specific mutations that increase blood vessel formation. Genetically modified stem cells have been shown to improve arteriogenesis in the hind limb of rabbits. For an artificial tissue engineered heart, such genetically engineered cells with higher collateral network formation property will be more resistant to ischemia.

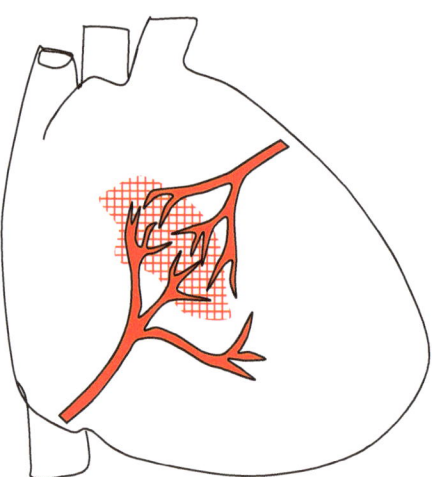

Extra collateral vascularized organs

Can all organs be vascularized with collateral redundant blood vessels for greater tolerance to blockage of arteries? Many organs do have some degree of collateral circulation but there are weak points that if blocked causes tissue death.

Extreme anastomosis - Mesh circulation

The output from the heart going to the rest of the body has a weak link. At least at some part of the path from heart to organ there is only one vessel. If it is blocked ischemia is inevitable. How about having say three or more sets of arteries coming out of the heart and all three being anastomosed every few centimeters. The result will be a robust circulatory system that is virtually block proof!

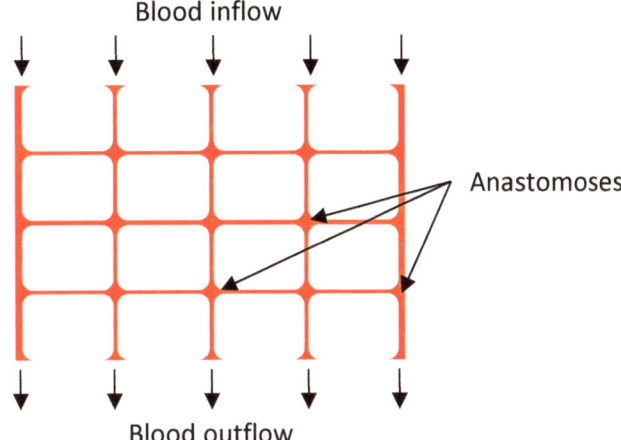

Low systemic resistance

Can we reduce the systemic vascular resistance of the arterial network permanently? Say by changing the architecture of capillaries to be more in number or in a different shape that allows better diffusion for same pressure? Reducing the systemic resistance reduces the load on the heart. With low resistance even a weak heart will probably be enough for survival. One way to increase nutrient throughput across the capillaries is active transport. Many of the molecules that passively diffuse across the capillary walls are powered by the pressure gradient created by the heart. If cells lining the capillary vessels can actively transport these using specific proteins, it will reduce the systolic pressure required from the heart. For oxygen, marine mammals seem to have better transport mechanism that lets them extract oxygen from blood at concentrations less than typically found in humans. Increased oxygen affinity of myoglobin and higher amounts of neuroglobins has been speculated as the reason for this.

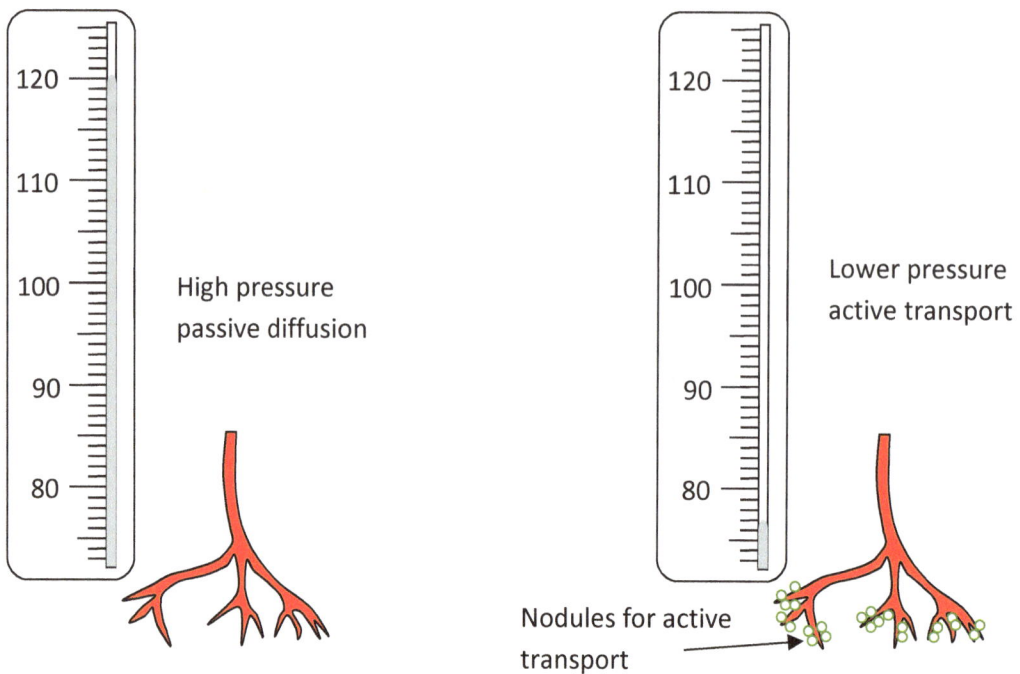

Valve leaflet count and valve performance

The heart valves in humans are mostly 2 or 3 leaflet types. Natural evolution improves an organ or tissue to a level that improves fitness to the environment but it does not mean it always produces the best possible organ for a given function. The heart valves also are pretty good but one can wonder if only 2 or 3 leaflets are the ideal number. What if number 7 was ideal? In fact there is considerable variation even among humans in the leaflet count. Some have 2 and others have 3 for the same type of valve. So, it's interesting to mathematically evaluate how many leaflets gives the best performing valve. The metric for performance is the amount of stenosis, regurgitation and durability. A cluster of valves with some sort of skeleton to hold them together can possibly make a good valve too.

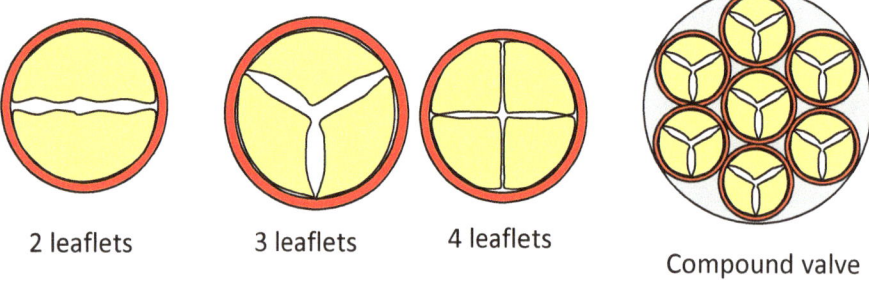

2 leaflets   3 leaflets   4 leaflets

Compound valve

Part 7 – Improvements to clinical practice

When it comes to publicity clinical practices do not carry as much weight as a high profile new stem cell therapy. Yet, clinical practices touch every person entering the hospital. This part discusses some methods that may improve the effectiveness of therapies in general.

Syringe array for stem cell delivery

One way to treat damaged tissue is to just plain inject stem cells into the damaged area such as heart muscle scarred by severe ischemia. The predominant approaches of delivery are epicardial patches, direct injection, arterial/venous catheter access and injection with micro-syringe. The repair capacity is directly related to the number of stem cells delivered. To cover a large area and deliver more cells an array of micro-syringes may be useful. So far, not many procedures use an array of micro-syringes.

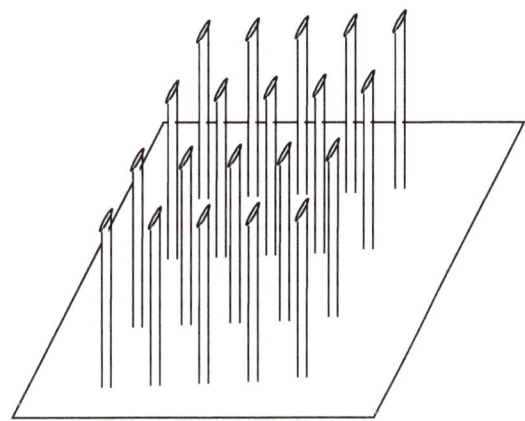

Unified graphical pathology model

Modern medical practice has access to vast amounts of literature covering broad areas. Significant amount of medical history also get generated for every individual. For any given patient entering the hospital, the physician is not able to take advantage of all these data when treating the patient. This is because the data is not presented in a user friendly and intuitive manner. Alongside, the evolution of computing technologies and communication systems make it possible to process the data with sophisticated algorithms. Compared to kids controlling TVs and video games with gestures, voice and touch screen, a typical physician seems to be underserved in connecting his brain with the external data. Quite simply, kids have moved to the smartphone era while physicians have not. Though touch screen interface is catching up, there does not seem to be a unified and easy to use representation of information. How about showing a 3D model of the person with parts of the body that are abnormal highlighted with markers? To get further information, the physician could zoom into the organ with abnormality. This zoom-in operation could be supported till the level of single nucleotide mutations. As the physician zooms into the area of the problem, the display could show the abnormal parameters super posed on the typical observed probability distribution of that parameter. Along the way, if the physician needs to know more about a particular test or report the display could switch to text mode and show the reports. All this information should be time stamped, i.e. the data associated with various systems and organs in the body should also be recorded periodically. This time information should by default be set to "now". If the physician chooses to see what these parameters were at a particular time in the past he could just move a touch screen "time" slider back. Even the best of the best physician cannot remember everything about human anatomy and physiology, let alone keep pace with latest developments in genetics, medical device technology and medical procedures. Conventionally, when a physician is at a loss to understand a particular piece of information about the patient he consults another physician expert in that area. Some features to supplement the knowledge of the physician can improve the speed of diagnostics and increase productivity. An "Assist" button can be added that helps the physician with possible differential diagnosis. Every diagnosis should be listed with the associated probability of being right so that the physician knows exactly how good the diagnosis is. As the physician medically "browses" through the patient graphics, memory aids can be provided with the "?" symbol. When clicked the "?" will bring up the definition of that piece of graphic and also provide further links to related literature. If the physician is interested in knowing how the disease could progress, he could move the "time" slider forward and the computer could be made to simulate the progression one of the most possible prediction for the given set of abnormalities. The future prediction should be clearly labeled with "PREDICTED". The physician should be given the option of looking at the prognosis of other diagnoses using radio buttons. Pieces of this idea can be found in the fields of health informatics specifically pathology informatics and clinical decision support systems. The idea of being able to zoom from whole body level to single DNA base level and being able to slide to past and future time seems not to be implemented in any system yet.

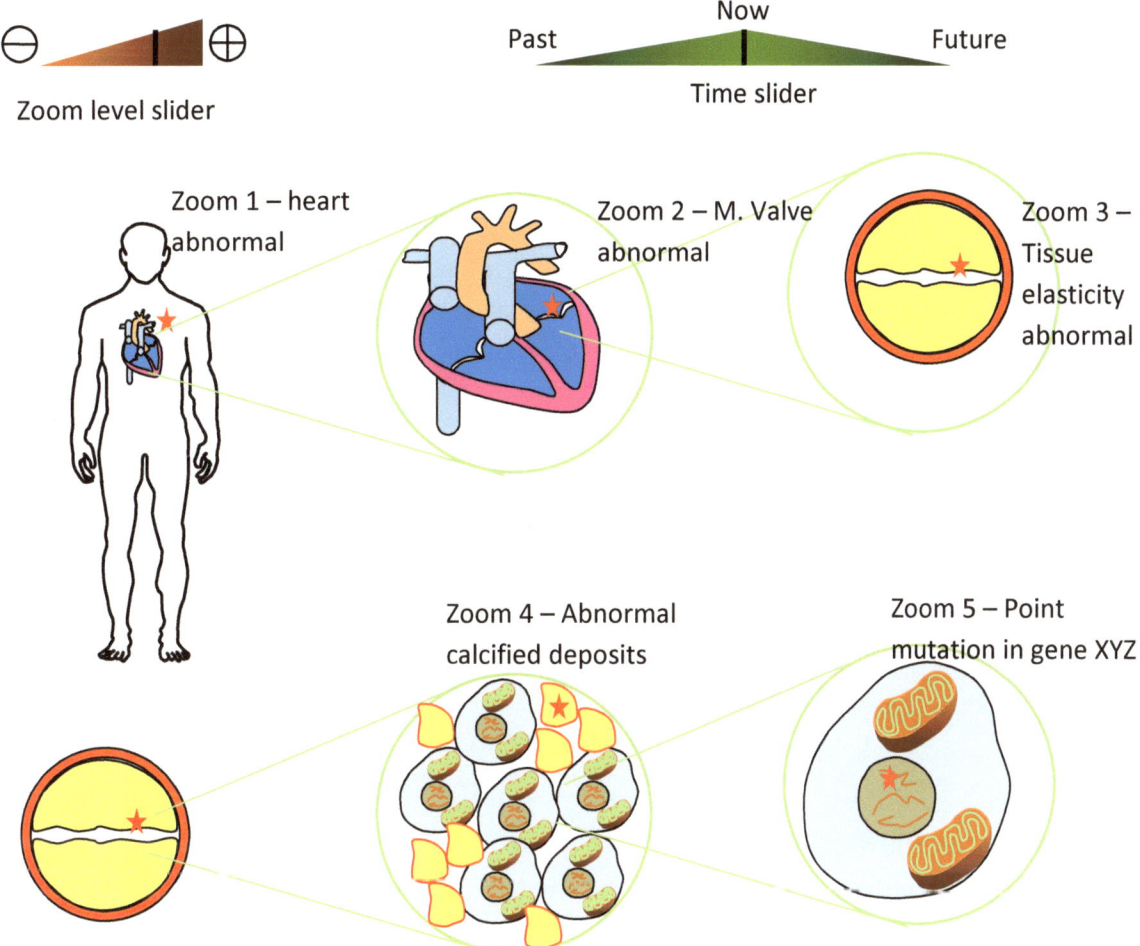

Zoom level slider

Past   Now   Future

Time slider

Zoom 1 – heart abnormal

Zoom 2 – M. Valve abnormal

Zoom 3 – Tissue elasticity abnormal

Zoom 4 – Abnormal calcified deposits

Zoom 5 – Point mutation in gene XYZ

Seamless Suturing

Although suturing is at least a few millennia old, no technique seems to provide instant closure of surgical incision with complete functional joining of both sides of the incision. Even fail-safe healing seems to be not yet achieved. Surgical wound dehiscing though occurring with low odds is still prevalent. Wouldn't it be useful to let a surgery patient just walk home immediately after the surgery with the confidence that the suture seam is as good as native tissue? Achieving seamless suture that is indistinguishable from uncut tissue will help eliminate extended hospital stays and reduce healthcare costs for surgery patients. Prior attempts have focused on using microfibers made of biocompatible materials, the use of glue, drug embedded sutures and mechanical contraptions. The fundamental problem of suturing is to return the cut tissue to its native functional state. The native state involves a proper amount of mechanical strength in the region, a proper amount of material exchange throughout the tissue and also proper nerve conduction. In theory, if every blood vessel, every extra cellular matrix filaments, nerves and lymphatic circulation is restored precisely the suture would be seamless and perfect. Practically, closing a cut with this much precision is impossible. It can certainly be made the theoretical ultimate target to compare a suture with. To begin with, suture thread thickness can be scaled down smaller and smaller. In the limit that the suture thread approaches the thickness of the native collagen fibers in the ECM, the suture will be indistinguishable from native tissue in terms of mechanical strength. As the thickness is scaled down, more and more fibers will be needed to close incisions necessitating a hand held robot suturing machine. Another difference between a suture and native tissue is cross linking, i.e. native tissue is more like a fabric with weaves in two or more directions whereas most sutures close incisions with threads mostly in one direction. Cross linking distributes mechanical stresses better than having threads in parallel without cross linking. For greater mechanical integration of the suture with the native tissue, it can be rooted farther away from the actual incision area. The ultimate suture, in my opinion, is possible using some sort of a hand held 3D printer that prints cells and ECM exactly where they should be right on the incision.

Conventional suture – Misaligned blood vessels, nerves, connective tissue

Seamless suture –Aligned blood vessels, nerves, connective tissue

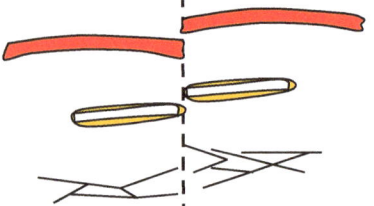 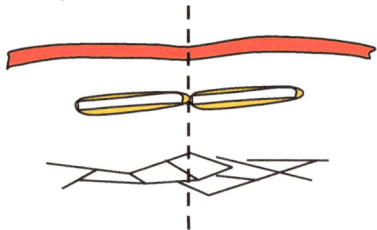

New clothing for medical professionals

While it's not very conclusive that medical professionals clothing carries a high risk of infection transmission, it does however pose a significant risk if not managed properly. Advances in synthetic biology, biochemistry and wearable technology can be used to improve the adherence to best hygiene practices. It can also be made doctor friendly. It is human to occasionally forget the laid out norms and use the clothing in an unapproved way. A professional could unknowingly wear a supposed to be sterile dress to a restroom and forget to change to a new one. Or, some small droplets containing pathogens could be stuck on the clothing without the knowledge of anyone. To allay some of these concerns, how about having some sort of electronic device embedded into the dress that reminds the doctor to change the dress every few hours, reminds to change the dress for every restroom visit and even reminds to change whenever visiting a higher risk contagious patient? RFID, Bluetooth low energy, IoT, NFC and WiFi may be usable for these purposes. Using latest developments in biotechnology, can we make color changing dress that changes color in the presence of body fluids like blood, saliva, mucous and other? In the extreme, it could be designed to detect even antibodies produced against a broad spectrum of pathogens and particles of pathogens. Yet, it should be fashionable to wear and let the professional personalize it to his tastes to improve the morale of the wearer. Another broad target for improvement is active disinfection. The current medical dresses are worn, accumulate dirt or germs and then washed clean. This cycle makes the dress uncomfortable for the wearer because he needs to change it often. It also leaves the wearer vulnerable to occasional lapses in hygiene. An actively sterilized dress continuously deactivating pathogens will extend the duration the dress can be worn. Some ways to actively sterilize the dress while being worn is to have micro heaters in the clothing that momentarily take the clothing to sufficiently high temperature. To avoid burning the wearer, an inner insulating lining can be provided. Another way is to imbue the clothing with biocide chemicals. While there are reports of using wearable technologies for monitoring healthcare workers and some reports of using clothing impregnated with biocides, no literature could be found for actively disinfected medical clothing. Also the reported improvements to medical clothing are not commonplace still.

Scrub with active disinfection
with micro-heaters

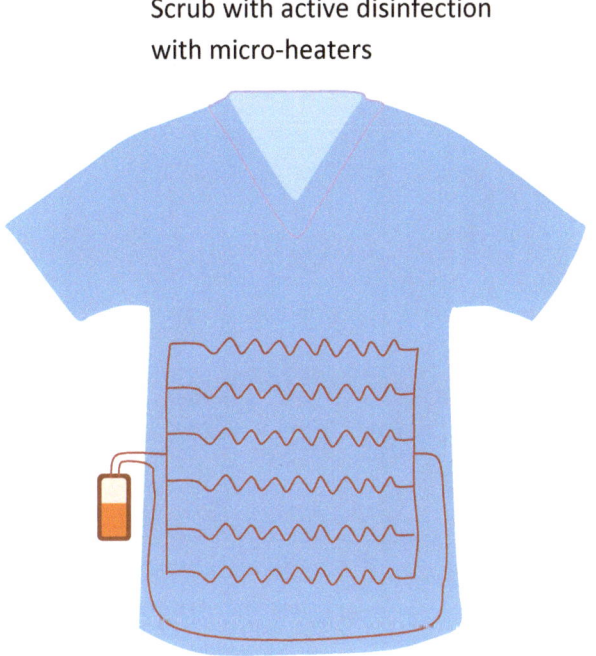

References

This book is essentially an offshoot from the book, "Handbook of Cardiac Anatomy, Physiology, and Devices", Second Edition, Paul A. Iaizzo et al., Springer Publications. For citations, only the first author is mentioned to keep it brief.

Part 1: Of the pump, by the pump and for the pump

Thinking Outside the Heart: Use of Engineered Cardiac Tissue for the Treatment of Chronic Deep Venous Insufficiency
Narine Sarvazyan, Journal of Cardio Vascular Pharmacology and Therapeutics, Feb 2014
http://www.ncbi.nlm.nih.gov/pubmed/24500906

http://cpt.sagepub.com/content/19/4/394
http://www.gizmag.com/mini-heart/31411
https://en.wikipedia.org/wiki/Pump
https://en.wikipedia.org/wiki/Diaphragm_pump
https://en.wikipedia.org/wiki/Cardiopulmonary_bypass

The Embryonic Vertebrate Heart Tube Is a Dynamic Suction Pump
Arian S. Forouhar et al., Science, 5 May 2006
http://www.sciencemag.org/content/312/5774/751.short
http://www.sciencemag.org/content/312/5774/751.short
https://en.wikipedia.org/wiki/Atrium_(heart)

Fish heart
http://esi.stanford.edu/circulation/circulation8.htm

Whole body pump
https://en.wikipedia.org/wiki/AutoPulse
http://www.ncbi.nlm.nih.gov/pmc/articles/PMC2998865/
http://www.lucas-cpr.com/en/lucas_cpr/lucas_cpr
http://circ.ahajournals.org/content/122/18_suppl_3/S720.full

Piezoelectric pump
https://en.wikipedia.org/wiki/Micropump
https://en.wikipedia.org/wiki/Piezoelectric_motor

MRI compatible pump
MRidium™ 3860+, IRADIMED corporation
http://www.iradimed.com/en-us/products/mridium3860.php

Targeting of Sodium Channel Blockers into Nociceptors to Produce Long-Duration Analgesia: A Systematic Study and Review
Roberson, DP et al. British Journal of Pharmacology 164.1 (2011): 48–58. PMC. Web. 13 Sept. 2015.
http://www.ncbi.nlm.nih.gov/pmc/articles/PMC3171859

Myocardial electrical impedance as a metric of completeness for radiofrequency ablation lesions,
Dumas, John Hicks, III, 2007, The University of North Carolina at Chapel Hill
Department Biomedical Engineering Joint Program, Ph.D. Dissertations & Theses
http://search.proquest.com/docview/304840266

Bioartificial Sinus Node Constructed via In Vivo Gene Transfer of an Engineered Pacemaker HCN Channel Reduces the Dependence on Electronic Pacemaker in a Sick-Sinus Syndrome Model
Hung-Fat Tse, MD, et al.,
http://online.liebertpub.com/doi/abs/10.1089/ten.TEB.2009.0352

Biological Pacemaker Engineered by Nonviral Gene Transfer in a Mouse Model of Complete Atrioventricular Block
Julien Piron, et al, Molecular Therapy, 2008
http://www.nature.com/mt/journal/v16/n12/abs/mt2008209a.html

Tissue Cardiomyoplasty Using Bioengineered Contractile Cardiomyocyte Sheets to Repair Damaged Myocardium: Their Integration with Recipient Myocardium
Miyagawa, Shigeru et al., Transplantation, Dec 2005,
https://www.researchgate.net/publication/7399078_Tissue_Cardiomyoplasty_Using_Bioengineered_Contractile_Cardiomyocyte_Sheets_to_Repair_Damaged_Myocardium_Their_Integration_with_Recipient_Myocardium

Bioartificial Sinus Node Constructed via In Vivo Gene Transfer of an Engineered Pacemaker HCN Channel Reduces the Dependence on Electronic Pacemaker in a Sick-Sinus Syndrome Model
Hung-Fat Tse, Circulation, 2006
http://circ.ahajournals.org/content/114/10/1000.short

Challenges in Cardiac Tissue Engineering
Gordana Vunjak-Novakovic et al, Tissue Engineering Part B: Reviews. April 2010
https://www.google.com/patents/US6238429

Neuron/cardiomyocyte coated contacts
https://www.google.com/patents/US4542752
http://www.google.com/patents/US20030211088

Part 3: Gasping for blood or drowning in blood – ischemia and reperfusion injury

Pre blocked bypass vessels
https://en.wikipedia.org/wiki/Xylitol

Replacement-parts
Ed Yong, TheScientist, Aug 2012
http://www.the-scientist.com/?articles.view/articleNo/32409/title/Replacement-Parts/

Animal organs for human transplantation: how close are we?
Marlon F. Levy, MD, Proc (Bayl Univ Med Cent)., 2000
http://www.ncbi.nlm.nih.gov/pmc/articles/PMC1312205/

https://en.wikipedia.org/wiki/Xenotransplantation

Bioengineered heart organiods
http://engineering.columbia.edu/bioengineered-heart-chambers-created

Normothermic cardiopulmonary bypass and myocardial cardioplegic protection for neonatal arterial switch operation
Philippe Pouard, European Journal of Cardio-Thoracic Surgery, 2006
http://ejcts.oxfordjournals.org/content/30/5/695.abstract

Physical and mechanical effects of cardioplegic injection on flow distribution and myocardial damage in hearts with normal coronary arteries.
Molina JE, J Thorac Cardiovasc Surg. 1989
http://europepmc.org/abstract/med/2542693

Blood Cardioplegia
Jurgen Martin, Multimedia manual of cardio thoracic surgery, 2006
http://mmcts.oxfordjournals.org/content/2006/1009/mmcts.2004.000745.full

Intermittent Cold-Blood Cardioplegia and Its Impact on Myocardial Acidosis during Coronary Bypass Surgery
Borowski A, Thorac Cardiovasc Surg, 2015
http://www.ncbi.nlm.nih.gov/pubmed/25423313

Chemical tweaks for cardioplegia
https://air.unimi.it/retrieve/handle/2434/4678/204097/lavoro%20n%C2%B0%207.pdf
https://en.wikipedia.org/wiki/Tetrodotoxin
https://en.wikipedia.org/wiki/Batrachotoxin
https://en.wikipedia.org/wiki/Maurotoxin

Autoimmune Channelopathies of the Nervous System
Kleopas A Kleopa, Curr Neuropharmacol. 2011
http://www.ncbi.nlm.nih.gov/pmc/articles/PMC3151600/

Cardiomyocyte death: mechanisms and translational implications
M Chiong, Cell Death and Disease, 2011
http://www.nature.com/cddis/journal/v2/n12/full/cddis2011130a.html

Mechanisms of cell death in acute myocardial infarction: pathophysiological implications for treatment
C. de Zwaan, Neth Heart J., Apr 2001
http://www.ncbi.nlm.nih.gov/pmc/articles/PMC2499566

Cell Death in the Pathogenesis of Heart Disease: Mechanisms and Significance
Russell S. Whelan, Annual Review of Physiology, March 2010
http://www.annualreviews.org/doi/full/10.1146/annurev.physiol.010908.163111

Antiapoptotic drugs: a therapautic strategy for the prevention of neurodegenerative diseases.
Sureda FX, Curr Pharm Des. 2011
http://www.ncbi.nlm.nih.gov/pubmed/21348832

Resveratrol suppresses apoptosis in intact human cardiac tissue - in vitro model simulating extracorporeal circulation.
Usta E, J Cardiovasc Surg (Torino), Jun 2011
http://www.ncbi.nlm.nih.gov/pubmed/21577194

Effect of hydrogen sulfide on myocardial protection in the setting of cardioplegia and cardiopulmonary bypass.
Osipov RM, Interact Cardiovasc Thorac Surg., Apr 2010
http://www.ncbi.nlm.nih.gov/pubmed/20051450

H2S Induces a Suspended Animation–Like State in Mice
Eric Blackstone, Science, April 2005
http://www.sciencemag.org/content/308/5721/518.short

Metabolic and functional effects of creatine phosphate in cardioplegic solution. Studies on rat hearts during and after normothermic ischemia.
Thelin S, Scand J Thorac Cardiovasc Surg. 1987
http://www.ncbi.nlm.nih.gov/pubmed/3589594

Intravenous ATP infusions can be safely administered in the home setting: a study in pre-terminal cancer patients
Sandra Beijer et al, Investigational New Drugs, Dec 2007
http://www.ncbi.nlm.nih.gov/pmc/articles/PMC2039853/

Nicotinamide adenine dinucleotide (NADH)--a new therapeutic approach to Parkinson's disease. Comparison of oral and parenteral application.
Birkmayer JG et al, Acta Neurol Scand Suppl. 1993
http://www.ncbi.nlm.nih.gov/pubmed/8101414

Adenosine and blood platelets
Hillary A. Johnston-Cox, Purinergic Signal. 2011 Sep
http://www.ncbi.nlm.nih.gov/pmc/articles/PMC3166992/

Part 4 - Rethinking energy delivery system

Alternative currency for biochemical energy
https://en.wikipedia.org/wiki/Explosophore
https://en.wikipedia.org/wiki/Alkaline_battery
https://en.wikipedia.org/wiki/Lithium_iron_phosphate_battery
https://en.wikipedia.org/wiki/Lithium-ion_battery

Part 5 – Genetic and cellular tweaks

Can ENCODE tell us how much junk DNA we carry in our genome?
Deng-Ke Niu, Biochemical and Biophysical Research Communications, Volume 430, Issue 4, 25 January 2013, Pages 1340–1343
http://www.sciencedirect.com/science/article/pii/S0006291X12024229

Enhanced multiplex genome engineering through co-operative oligonucleotide co-selection
Peter A. Carr, Nucleic Acids Res. 2012 Sep; 40(17): e132.
http://www.ncbi.nlm.nih.gov/pmc/articles/PMC3458525/

Minimization of the Escherichia coli genome using the Tn5-targeted Cre/loxP excision system.
Yu BJ, Methods Mol Biol. 2008
http://www.ncbi.nlm.nih.gov/pubmed/18392973

Analysis of a genome-wide set of gene deletions in the fission yeast Schizosaccharomyces pombe
Dong-Uk Kim, Nature Biotechnology 28, 617–623 (2010) doi:10.1038/nbt.1628
http://www.nature.com/nbt/journal/v28/n6/abs/nbt.1628.html

CIRM grant for personalized embryonic stem cell like cells
https://www.cirm.ca.gov/our-progress/awards/formation-personalized-embryonic-stem-cells-vitro-epigenetic-cell-reprogramming

Toward eliminating HLA class I expression to generate universal cells from allogeneic donors
Hiroki Torikai, Blood, Aug 2013
http://www.ncbi.nlm.nih.gov/pmc/articles/PMC3750336

Comparison of the molecular profiles of human embryonic and induced pluripotent stem cells of isogenic origin
Barbara S. Mallona et al., Stem Cell Research, Volume 12, Issue 2, March 2014, Pages 376–386
http://www.sciencedirect.com/science/article/pii/S1873506113001724

Analysis of long-lived C. elegans daf-2 mutants using serial analysis of gene expression
Julius Halaschek-Wiener, Genome Res. 2005 May; 15(5): 603–615.
http://www.ncbi.nlm.nih.gov/pmc/articles/PMC1088289/

Tweaks proposed by SENS foundation
http://www.sens.org/research/introduction-to-sens-research

Apoptosis pathways and genes
http://www.genome.jp/dbget-bin/www_bget?hsa04210

Gene search for cell motility produces more than 900 genes
http://www.ncbi.nlm.nih.gov/gene/?term=homo+sapien+cell+motility

The role of mitochondria in aging
Ana Bratic et al, The journal of clinical investigation, March 2013
http://www.jci.org/articles/view/64125

The evolving role of mitochondria in metabolism, Transcriptional integration of mitochondrial biogenesis
Richard C. Scarpulla, Trends in Endocrinology & Metabolism, Volume 23, Issue 9, September 2012
http://www.sciencedirect.com/science/article/pii/S1043276012001075

Mitochondrial longevity pathways
M.H. Vendelbo, Biochimica et Biophysica Acta (BBA) - Molecular Cell Research Volume 1813, Issue 4, April 2011, Pages 634–644
http://www.sciencedirect.com/science/article/pii/S0167488911000371

Genetically enhanced hemoglobin
https://en.wikipedia.org/wiki/High-altitude_adaptation

Evolutionary and functional insights into the mechanism underlying high-altitude adaptation of deer mouse hemoglobin
Jay F. Storza, et al
http://www.pnas.org/content/106/34/14450.long

Dynamic Factors Affecting Gaseous Ligand Binding in an Artificial Oxygen Transport Protein
Lei Zhang, Biochemistry 2013 52 (3), 447-455
http://pubs.acs.org/doi/abs/10.1021/bi301066z

https://en.wikipedia.org/wiki/Hemocyanin

https://en.wikipedia.org/wiki/Cytoglobin
https://en.wikipedia.org/wiki/Neuroglobin
https://en.wikipedia.org/wiki/Myoglobin
https://en.wikipedia.org/wiki/Hemerythrin
https://en.wikipedia.org/wiki/Dioxygenase
https://en.wikipedia.org/wiki/Dioxygenase#Cofactor-independent_dioxygenases
https://en.wikipedia.org/wiki/Ligand
https://en.wikipedia.org/wiki/Ethylenediaminetetraacetic_acid

Comprehensive Coordination Chemistry II
D.M. Kurtz Jr., 2003, Pages 229–260, Volume 8: Bio-coordination Chemistry 8.10 – Dioxygen-binding Proteins
http://www.sciencedirect.com/science/article/pii/B0080437486081718

Hemolysis and Acute Kidney Failure.
Qian, Qi et al., American Journal of Kidney Diseases 56.4 (2010): 780–784. PMC. Web. 27 Nov. 2015.
http://www.ncbi.nlm.nih.gov/pmc/articles/PMC3282484/

Genetically Engineered Hemoglobin Brings Artificial Blood A Step Closer
November 7, 2000, American Chemical Society
http://www.sciencedaily.com/releases/2000/11/001107070452.htm

Genetically enhanced RBC
https://en.wikipedia.org/wiki/Plasmodium_falciparum
https://en.wikipedia.org/wiki/Red_blood_cell#/media/File:Erythrocytes_in_vertebrates.jpg

Temperature effects on Ca2+ cycling in scombrid cardiomyocytes: a phylogenetic comparison
Gina L. J. Galli, J Exp Biology, Apr 2011
http://onlinelibrary.wiley.com/doi/10.1113/jphysiol.2013.261198/full

Ionic mechanisms limiting cardiac repolarization reserve in humans compared to dogs
Norbert Jost, The Journal of Physiology, Sep 2013
http://www.ncbi.nlm.nih.gov/pubmed/23878377

Bioengineering Human Myocardium on Native Extracellular Matrix
Jacques P. Guyette et al., Circulation Research. 2016
http://circres.ahajournals.org/content/118/1/56.short

Iterative operator splitting method for capillary formation model in tumor angiogenesis problem: Analysis and application
Nurcan Gücüyenen, International Journal for Numerical Methods in Biomedical Engineering, Volume 27, Issue 11, pages 1740–1750, November 2011
http://onlinelibrary.wiley.com/doi/10.1002/cnm.1435/abstract

Part 6 – Optimizing circulation

Species variation in the coronary collateral circulation during regional myocardial ischaemia: a critical determinant of the rate of evolution and extent of myocardial infarction
Miles P Maxwell et al, Cardiovascular Research, Oct 1987
http://cardiovascres.oxfordjournals.org/content/21/10/737.short

Genetically modified human placenta derived mesenchymal stem cells with FGF 2 and PDGF BB enhance neovascularization in a model of hindlimb ischemia.
Yin, T. et al, Molecular Medicine Reports, 12, 5093-5099.
http://dx.doi.org/10.3892/mmr.2015.4089

Fibroblast growth factors in cardiovascular disease: The emerging role of FGF21
Eleni M. Domouzoglou et al, American Journal of Physiology - Heart and Circulatory Physiology Sep 2015
http://ajpheart.physiology.org/content/309/6/H1029
https://en.wikipedia.org/wiki/Circle_of_Willis

Myoglobin oxygen affinity in aquatic and terrestrial birds and mammals
Traver J. Wright, Randall W. Davis, Journal of Experimental Biology 2015
http://www.ncbi.nlm.nih.gov/pubmed/25987728

Running, swimming and diving modifies neuroprotecting globins in the mammalian brain
Terrie M Williams et al, Proc. R. Soc. B 2008
http://www.ncbi.nlm.nih.gov/pubmed/18089537

https://en.wikipedia.org/wiki/Pinocytosis
https://en.wikipedia.org/wiki/Receptor-mediated_endocytosis
https://en.wikipedia.org/wiki/Paracellular_transport
https://en.wikipedia.org/wiki/Transcellular_transport

The bicuspid aortic valve: an integrated phenotypic classification of leaflet morphology and aortic root shape
B M Schaefer, Heart 2008
http://www.ncbi.nlm.nih.gov/pubmed/18308868

Review in Basic and Applied Anatomy Anatomic variations of the cardiac valves and papillary muscles of the right heart
Theodoros Xanthos, Italian Journal of Anatomy and Embryology, 2011
www.fupress.net/index.php/ijae/article/viewFile/10339/9524Cachedcles

Reptilian cardiovascular anatomy and physiology: evaluation and monitoring
Ryan S. De Voe, DVM, MSpVM, DACZM, DABVP CVC In San Diego Proceedings

http://veterinarycalendar.dvm360.com/reptilian-cardiovascular-anatomy-and-physiology-evaluation-and-monitoring-proceedings?rel=canonical
http://ir.uiowa.edu/cgi/viewcontent.cgi?article=4637&context=etd

Part 7 – Improvements to clinical practice

Needle array for stem cell delivery
http://www.cathlabdigest.com/articles/Options-Stem-Cell-Delivery-Cardiology

Mercator Medsystems microsyringe
http://www.mercatormed.com/product3.html

Epicardial delivery of collagen patches with adipose-derived stem cells in rat and minipig models of chronic myocardial infarction.
Araña M, Biomaterials, Jan 2014
http://www.ncbi.nlm.nih.gov/pubmed/24119456

Cerner Corporation – Physician practice
http://www.cerncr.com/solutions/Physician_Practice/

Sunquest Information Systems – Sunquest VUE product
http://www.sunquestinfo.com/products-solutions/anatomic-pathology/sunquest-vue-the-diagnosticians-workstation

LogicNets, Inc.
http://medical.logicnets.com/
https://en.wikipedia.org/wiki/Clinical_decision_support_system

Google maps for the human BODY: Researchers unveil interactive images that let them zoom to cellular level
Mark Prigg for Dailymail.com, 30 March 2015
http://www.dailymail.co.uk/sciencetech/article-3018500/Google-maps-human-BODY-Researchers-unveil-interactive-images-let-zoom-cellular-level.html
https://en.wikipedia.org/wiki/Surgical_suture
https://en.wikipedia.org/wiki/Wound_dehiscence

Trends in postcoronary artery bypass graft sternal wound dehiscence in a provincial population.
Doherty C1, Plast Surg (Oakv). 2014 Fall;22(3):196-200.
http://www.ncbi.nlm.nih.gov/pubmed/25332650

Closure of long surgical incisions with a new formulation of 2-octylcyanoacrylate tissue adhesive versus commercially available methods
Phillip N.V. Blondeel, M.D., American Journal of Surgery, Sep 2004
http://www.americanjournalofsurgery.com/article/S0002-9610(04)00215-6/abstract

Development of drug loaded microfiber sutures for ophthalmic application
Peter McDonnell, Johns Hopkins University
http://jhu.technologypublisher.com/technology/17130

3D Bioprinting of Tissues and Organs
Sean V Murphy & Anthony Atala, Nature Biotechnology 32, 773–785 (2014)
http://www.nature.com/nbt/journal/v32/n8/full/nbt.2958.html

Toward the Development of a Hand-Held Surgical Robot for Laparoscopy," in Mechatronics, Zahraee, A.H.et al, IEEE/ASME Transactions on , vol.15, no.6, pp.853-861, Dec. 2010
http://ieeexplore.ieee.org/xpl/login.jsp?tp=&arnumber=5523951

Role of healthcare apparel and other healthcare textiles in the transmission of pathogens: a review of the literature
Mitchell, M. Spencer et al, Journal of Hospital Infection, Volume 90, Issue 4, August 2015, Pages 285-292, ISSN 0195-6701
http://dx.doi.org/10.1016/j.jhin.2015.02.017
http://www.nytimes.com/2008/09/23/health/23well.html?_r=0
https://en.wikipedia.org/wiki/Scrubs_(clothing)
https://en.wikipedia.org/wiki/Radio-frequency_identification
https://en.wikipedia.org/wiki/Wi-Fi
https://en.wikipedia.org/wiki/Near_field_communication
https://en.wikipedia.org/wiki/Bluetooth

Measuring Healthcare Worker Hand Hygiene Activity: Current Practices and Emerging Technologies
John M Boyce (2011), Infection Control & Hospital Epidemiology, 32, pp 1016-1028.
http://journals.cambridge.org/action/displayAbstract?aid=9374438

The feasibility of an automated monitoring system to improve nurses' hand hygiene
Levchenko, Alexander I. et al., International Journal of Medical Informatics , Volume 80 , Issue 8 , 596 – 603
http://www.ijmijournal.com/article/S1386-5056(11)00104-3/abstract

Deoxyribozyme Cascade for Visual Detection of Bacterial RNA
Dr. Yulia V. Gerasimova1 et al, ChemBioChem Volume 14, Issue 16, pg 2087–2090, November 4, 2013
http://onlinelibrary.wiley.com/doi/10.1002/cbic.201300471/full

Reduction of healthcare-associated infections in a long-term care brain injury ward by replacing regular linens with biocidal copper oxide impregnated linens
Lazary, A. et al., International Journal of Infectious Diseases , Volume 24 , 23 – 29
http://www.ijidonline.com/article/S1201-9712(14)00059-9/fulltext

https://en.wikipedia.org/wiki/Electric_blanket

www.ingramcontent.com/pod-product-compliance
Lightning Source LLC
Chambersburg PA
CBHW040744200526
45159CB00023B/1678